T0136848

Lecture Notes in Energy

Volume 69

Lecture Notes in Energy (LNE) is a series that reports on new developments in the study of energy: from science and engineering to the analysis of energy policy. The series' scope includes but is not limited to, renewable and green energy, nuclear, fossil fuels and carbon capture, energy systems, energy storage and harvesting, batteries and fuel cells, power systems, energy efficiency, energy in buildings, energy policy, as well as energy-related topics in economics, management and transportation. Books published in LNE are original and timely and bridge between advanced textbooks and the forefront of research. Readers of LNE include postgraduate students and non-specialist researchers wishing to gain an accessible introduction to a field of research as well as professionals and researchers with a need for an up-to-date reference book on a well-defined topic. The series publishes single- and multi-authored volumes as well as advanced textbooks.

More information about this series at http://www.springer.com/series/8874

Janet Nagel

Optimization of Energy Supply Systems

Modelling, Programming and Analysis

 Springer

Janet Nagel
Berlin, Germany

ISSN 2195-1284   ISSN 2195-1292   (electronic)
Lecture Notes in Energy
ISBN 978-3-030-07180-6   ISBN 978-3-319-96355-6   (eBook)
https://doi.org/10.1007/978-3-319-96355-6

This Springer imprint is published by the registered company Springer Nature Switzerland AG
The registered company address is: Gewerbestrasse 11, 6330 Cham, Switzerland

Preface

This book continues my strong commitment to energy supply as a topic. It represents a return to my roots in energy supply modeling and optimization of energy supply systems. The first time I worked on the topic of optimizing energy supply systems was more than 20 years ago. Even then, the models selected differed enormously, some favored heuristic models, some focused on agency, and others used classic linear optimization models. In the context of the energy transition in Germany, this topic has garnered significantly more interest than before. Around the globe, countless questions about how to solve the energy supply problem for a sustainable future for all urgently await an answer.

Almost by accident, I found out about an exciting new project, or rather, an exciting new system of program in this field, the *Open Energy System Modelling Framework* (OEMEF). This project, or rather, this tool, can make an important contribution to answering these myriad questions on optimizing energy supply and thus can contribute to making the German and the global energy transition a reality. OEMOF provides a generic approach for the modeling and optimization of energy supply systems. Generic approaches, i.e., approaches that work in many different contexts, are a key feature of object-oriented programming. The generic concept of OEMOF is reflected in the generic and recurring description of classes and functions. This allows OEMOF to be applied to very different types of problems.

I had started working on databases more than 15 years ago and was particularly interested in generic approaches. My idea was that having to reprogram databases again and again and having to continually adapt data models seemed inefficient. I therefore looked into creating a generic data model. More information about publications on this topic can be found at http://www.risa.eu/de/safetyanalyses/publications.php.

The OEMOF project is also inspired by the vision that those interested in optimizing energy supply systems will no longer have to create all their models from scratch but can make use of this wonderful resource to concentrate on reflecting the specific aspects of their optimization problem. At the moment, this system of program is still in its development stage. As this is an open source project, this means that you are all invited to improve its design and use your knowledge to further its development.

OEMOF is continually being improved and redeveloped by the OEMOF community. Even while I was writing this book, OEMOF development moved on to a new stage by issuing Release v0.2.0. As this new version was released just before the book was completed, it was not possible to include user feedback on how this new Release handles here.

This book is designed to be a first step toward bringing the OEMOF user and developer community together and introducing the tool to potential modelers. When I started working with the tool I wanted to understand and to pass on to my readers how the programming of the modeling software was carried out. This is the reason why this book gives several source code excerpts and provides a number of examples of OEMOF applications.

The OEMOF developer community, and in particular those who have started this project and continue to work on it, will have their very own views on OEMOF and on how to write about it. After all, they have developed the tool and gave it life. For this reason, this is a book *about* OEMOF and not a book *by* OEMOF, i.e., written by the OEMOF community. Nevertheless, I am convinced that this book will be a great help for getting started with modeling in OEMOF. My own work researching at university level has shown me that modelers who have never before worked on optimization questions can find it hard to understand how to work with this tool. Therefore, I believe that this book, though it cannot answer all your questions, will help get you into OEMOF.

The quality of books depends on the expert knowledge and range of topics contained in them; this is something that I have made use of here as in all my previous books. The experts that have to be acknowledged here are first and foremost the OEMOF community. They have helped me gain a deep insight into the program by providing workshops and discussions and by commenting parts of the manuscript for this book. I would like to thank them very much for their support.

I would also like to give a special thanks to Dr. Stöhr from B.A.U.M. Consult GmbH for his active support of this book project. By means of discussions and constructive exchanges, his views and insights have fed into the topic of optimizing energy supply systems, which is so prominent in the book. Dr. Stöhr's love of detail and his interest and enthusiasm for portraying mathematical aspects as clearly and precisely as possible have inspired me to delve more deeply into this subject. As we all know how precious time is in our professional lives and how busy Dr. Stöhrs diary, I am even more grateful for how generous he has been with his time and support in passing on his expert knowledge.

I had always planned to publish this book in English. This was made possible with the assistance of Prof. Dr. Verena Jung of the AKAD University, professor of linguistics and translation. Her expertise in translating this book into English makes it available to a wider readership. I would like to thank her for completing this important task.

Scenario building plays a vital role in developing energy models. Within the working group "Internet of Energy" of the BDI (The Federation of German Industries), especially with Detlef Schumann, BridgingIT GmbH, I discussed important methodical approaches for this chapter. I would like to thank Mr. Schuhmann for these inspiring talks and impulses.

As this book is largely concerned with the Python programming language, I have sought out the expert support of Henrik Fahlke, developer and Python specialist. He answered all my questions on Python and helped me write my own little example in code. I would like to thank him very much for his time and his support.

I would also like to thank my faithful publishing colleague Dr. Silvia Porstmann, CEO of Seramun Diagnostica GmbH, for her support and work on this manuscript. In particular, her proofreading of the English translation has been invaluable. I would like to thank her for her keen interest and her commitment, with which she also supported my previous two book projects.

A book project can only thrive if it is supported by its publisher. Springer Publications have been a very competent and supportive partner in this endeavor. I would like to thank all Springer Publications employees involved in this project.

Last but not least, my very particular thanks must go to my family. Not only have they actively supported me in this book project, but more importantly, they took the many hours that I worked on this project during our holidays or on the weekend in their stride. Without your support, none of these projects would have been possible. Thank you so much!

Finally, I would like to wish all readers of this book, energy modelers, and future OEMOF developers an enjoyable introduction to this wonderful tool. Together we will get the energy transition off the ground!

Berlin, Germany Janet Nagel
April 2018

Contents

Chapter 1
The OEMOF Project

The world is coming closer together in the issue of securing future energy supply. Previously, oil and gas were mainly imported from other countries or exported to other countries. With the expansion of renewable energies new ideas have come into the world. The power generation and distribution no longer concerns one country alone. Instead, there are dependencies between countries. One example is Germany, as a country that shares borders with a number of other European countries. There is a brisk trade in energy between countries. The integration of economies within Europe is increasing and becoming more and more complex. The focus is always on the question of how the energy supply can be ensured while at the same time fulfilling certain conditions, such as low CO_2 emissions.

However, not only transnational issues are of interest when looking at energy development in the future. Even for a single country, such as Germany, or even a single German state, such as Bavaria, or even a region within a state or the area covered by one local energy supplier, the question of the future shape of energy supply is of crucial importance. Figure 1.1 shows the complexity of relations within the energy supply sector.

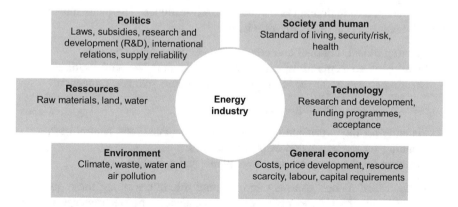

Fig. 1.1 Dependencies and influences in the energy sector

© Springer International Publishing AG, part of Springer Nature 2019
J. Nagel, *Optimization of Energy Supply Systems*, Lecture Notes in Energy 69,
https://doi.org/10.1007/978-3-319-96355-6_1

Further information about the energy supply situation in Germany and global issues and technological solutions can be found in Nagel (2017).

In Germany municipalities or private energy suppliers or even residents of an area or else a company or a group of companies can agree to run their own energy supply system and take on the responsibility for an energy supply system. If the energy supply system is a new plant or a new energy supply location, the power plant is designed and planned by engineers.

Sometimes energy supply systems are considered trans-regionally, for example the whole energy supply structure of Germany or France. The issues concerned for such scenarios are far-reaching and global in nature. The most important factor is not to get a concrete result for an optimization function, but to establish a general framework and future contexts for an energy supply system, such as trying to create a supply system with minimal CO_2 emissions.

There are several key objectives when developing energy supply systems:

- The energy needs of the population must be fulfilled by ensuring an adequate supply.
- The energy supply system must administer the energy distribution efficiently in such a way as to minimize the cost to the economy.
- The energy supply must be sustainable and ensure responsible treatment of natural resources.
- The energy supply is to have little or no harmful effect on the health and the lifestyle of the population.
- All resources are to be used efficiently in order to ensure that future generations will be able to lead the lives they want.

These objectives are in conflict with each other, and these conflicts must be resolved and managed. This means that decisions need to be taken and plans must be set up.

These plans must give answers to the question of the shape of the future energy system. But the plans will also have to gage the effects on the current energy supply system, the effect on the market and on the environment. The plans must also engage globally with stakeholders such as politicians, supplier and consumer associations, consultation agencies and research institutes. The plans must include the following aspects of energy provision:

- Creating global, regional and community energy supply concepts.
- Developing concepts for efficient energy use in buildings and factories.
- Creating concepts for reducing pollutant emissions (climate protection).

Further required planning aspects include:

- Planning for the expansion of supply networks (electricity, natural gas).
- Planning for the expansion of power stations.
- Planning for the operation of power plants and refineries.

Mathematical models can play an important role in answering these questions. The model must be able to map the interdependencies between energy demand, power generation, energy storage, markets, pricing, and the impact on the environment. The issues can be characterized as being multidimensional, highly complex, having long-lasting consequences, and being highly volatile. The many conflicting objectives further complicate the issue.

In order to be able to accommodate the various issues and the conflicting objectives, energy supply optimization planning normally uses highly individualized versions of different programming and modeling languages, such as GAMS (General Algebraic Modeling System). These models have been in use for several decades. Further information can be found in (Forum 1999; König 1977; Pfaffenberg 2004; Walbeck 1988; Schultz 2008; Kallrath 2013).

The OEMOF project offers a new approach. The abbreviation OEMOF stands for 'open energy modeling framework'. OEMOF provides a toolbox that can be used for the modeling of energy systems. OEMOF can be used to generate energy feed time series, load profiles as well as entire energy models. A model library for the development of energy models is available at oemof.solph. This model library is the centerpiece for future modelers. The library also contains the mathematical equations required for modeling all components of the energy supply system. Using the model library, modelers can generate equations mapping objectives, such as the energy supply costs. As customary for optimization functions, a minimum or maximum value is computed. The equations provided at oemof.soph always provide minimum values, such as the lowest cost. Costs can be calculated as fixed costs (time-dependent costs) or as variable costs (quantity dependent). Similarly, costs for pollution or other parameters can be defined. Investment costs and depreciation for components and facilities or buildings can be assigned. These models are all cost-based modelings. These models help energy development planners decide between different options based on their associated cost. But it is also possible to use the toolbox and the examples at oemof.solph to calculate the emissions minimum. This means that OMOEF can help to generate a climate protection energy model. In a cost-based model, emissions can be entered as one of the restriction parameters.

One important characteristic of OEMOF is the fact that the system already provides all generic features. Heat and power generation systems, energy distribution costs, pricing, and many other key aspects are already mathematically described in this model and in the model library and can be arranged by the modeler in the desired combination relevant for answering virtually any type of question. Modelers can thus generate a great variety of different models. Modelers can focus on adjusting the relevant parameters for their question, without having to engage with creating the equations themselves. One further important characteristic of OMOEF is the fact that it is available free of charge (open source). As OEMOF is an open source modeling toolbox, users can engage with it and refine it. Furthermore, OEMOF was conceived as a collaborative project, which means that all further developments can be fed back to the modeling community. The programming system was created based on a specific approach which will be introduced in the next section after a brief introduction to mathematical modeling.

OEMOF is spearheaded by the Reiner Lemoine Institute and the Center for Sustainable Energy Systems (ZNES). The ZNES is a joint institution of Flensburg

University of Applied Sciences and the Europe University at Flensburg. They initiated the project together and they developed the first publicly available version of OEMOF. Today a number of scientists jointly work on this project such as members of the Technische Universität Berlin, Germany and stakeholders and users such as the B.A.U.M. group collaborate on this project. A complete list of all project collaborators and users of OEMOF can be found on the website of the OEMOF initiative (https://wiki.openmod-initiative.org/wiki/Special:ListUsers).

At the end of 2017, a new version of OEMOF was issued, Release v0.2.0. The discussion and explanations in this book all refer to this release.

1.1 Methods for Developing Strategies and Options

The energy transition is in full progress and no one really knows what the energy supply structure will look like in 2050. But in order to take decisions for the future and in order to set the right course, key influencing factors must be identified and predictions must be made. Energy models are a key factor in forecasting. By means of a model, the current energy supply system can be analyzed, strategies can be developed and actions taken. Each model is based on a question. One such question could be: What is the pricing and payment structure that will enable us to produce biomethane and biogas from organic residues and waste? Another question might be: What is the framework that will enable rural regions to use biogenic materials for heat generation (Nagel 1998)?

Energy models can be used to map interdependencies between different agents within an energy system and to reflect different economic performance indicators as well as different environmental factors. These interdependencies are mapped and computed by means of mathematical equations or rather, by a system of mathematical equations. Together, these equations constitute the energy supply model.

Such a model is based on a vision of the future. In order to forecast future energy structures, narrative scenarios are created that describe the forecast as a storyline (Krutzler 2016). Within these scenarios, specific figures are predicted, such as 'In the year 2050, 80% of the world's population will drive an electric vehicle'. Often, more than one scenario is developed. As part of the future forecast, different scenario alternatives are presented. In many cases, three different scenarios are defined:

- Best Case Scenario
  This scenario presents a forecast for a development where all strategies have been successfully implemented.
- Worst Case Scenario
  This scenario presents a forecast for a development in which the worst fears for the future have come true.
- Realistic scenario
  This scenario reflects the assumption that there will be divergence from the ideal scenario.

Scenarios examine possible developments and create an image of the future. But they do not make a statement about the likelihood of their coming true, they are not a prediction. Rather, they describe one possible version of developments of parameters that cannot be influenced by the decision-makers. These scenarios are used in the energy sector to reflect a stringent development in key energy parameters. For example, one of these key parameters could be energy consumption. This key parameter could be defined as requiring the expansion of renewable energy use and a reduction of CO_2 emissions.

Based on each of these scenarios the different strategies and actions required for enabling this development can be planned.

In addition to scenarios, other methods and techniques for developing strategies and actions are storylines and models. All three methods differ with regard to their forecasting potential in terms of quantity and quality (cf. Fig. 1.2).

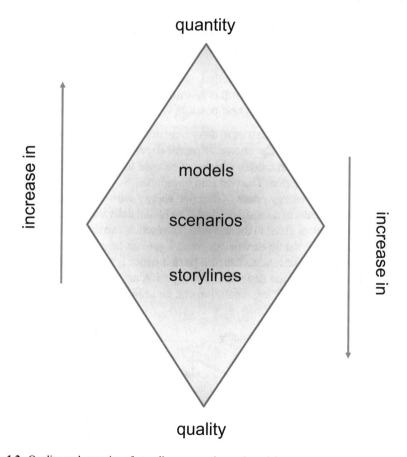

Fig. 1.2 Quality and quantity of storylines, scenarios and models

Different models can be used for analyzing energy supply systems. The following different types of models are used:

– Forecasting models
    Their task is to predict the future. They are used in the energy sector to forecast sunshine hours and sunshine intensity.
– Analytical models
    Different parameters such as population figures or gas prices are analyzed with regard to their influence on the energy system under review in order to gage and determine their relevance.
– Scenario models
    Here, the focus is on the creation of different rival scenarios, in order to represent several different possible future developments and their possible implications.
– Simulation models
    These models reflect reality. For example, the motion of a wind turbine can be simulated in order to calculate its energy yield.
– Optimization models
    In these models, an objective function, such as a cost equation, is defined. The minimum or maximum value for this function is to be determined. By means of an algorithm, the lowest or highest possible value for this function is calculated.

All of these models allow the user to determine strategies and actions for the future based on their forecasting. The choice of model depends on the research question. OEMOF is an optimization model, which can be used to determine the lowest possible value for a cost function. The cost is to be reduced. Several different parameters can be varied, such as energy prices of different energy sources in order to determine their influence by means of a sensitivity test. The model delivers a vision of the future based on the parameters given in the scenario. The models can thus predict how many wind parks have to be set up or whether natural gas can be an economically viable energy supply option. The models can also predict which types of biogas plants, with or without heat cogeneration and with what capacity, are required.

The method for determining a solution can be seen in Fig. 1.3.

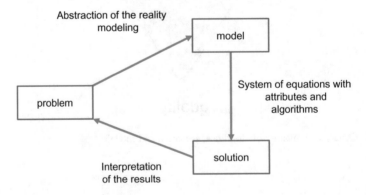

Fig. 1.3 Formal approach to a problem

1.2 What Is OEMOF Based on?

OEMOF and in particular the oemof.solph library models energy sectors such as heat, cold and electricity as well as the corresponding technologies available on the market, such as CHP plants or power to gas by means of transforming electricity to hydrogen or methane by means of electrolysis. The energy implications of electric mobility are also described. This allows users to look at energy supply across all energy sectors and to study sector interconnections and interdependencies.

Oemof.solph can act as a model generator. Modelers can use it to create one energy supply model, either a linear or a mixed-integer optimization model, combined from the different components available at the model library.

The model library also allows modelers to link and combine different regions and map them in one model. Thus it is possible to use oemof.solph to map the 16 German states separately and then to combine the different analyses into one model.

The timeline of energy supply is also of key importance for gaging energy supply security, as this is one of the key objectives of energy supply systems. The objective of energy security requires an energy system that can cover the energy demand at any time. The smaller the chosen time increments of the model, the more precise the model can map the energy supply across a defined time period. Oemof.solph allows the free choice of time increment to be mapped. This enables the modeler to choose the time increment based on the data available.

Furthermore, the model library oemof.solph is modular in structure. What does this mean? Each constituent, i.e., each element, such as a CHP plant or a gas supply grid, corresponds to one module, in which this element is described either as a linear equation or as a mixed-integer linear equation system. Each constituent can be further defined by a certain number of parameters, such as degree of efficiency or plant capacity. It is part of the task of the modelers to define the values for these parameters. These constituents are programmed generically in the model library at oemof.solph. In this way, the very different constituents can all be mapped. Some examples for constituents are listed below:

– Energy generation systems, such as cogeneration plants, wind turbines, solar arrays, diesel-generators, CHP plants, heating plants.
– Facilities for the transport of energy, such as electricity grids, gas grids, or heat pipelines.
– Different types of consumers, such as cities, buildings, companies and industrial manufacturers.
– Fuels, such as oil, gas or wood.
– Exports and imports, such as gas import, coal mining, or electricity export.

The modular system is based on clear demarcations between separate systems. One system represents all elements that are interdependent. One example of a separate energy system is the city of San Diego on the West Coast of the United States. All energy constituents are part of this energy system, starting from natural energy sources such as wood, wind or solar energy via the transformation of energy by means of gas turbines, via the energy distribution by means of an electricity grid and the storage in a battery for the final consumer, e.g., an individual household, a machine, a technical device or a utility provider.

What is OEMOF's generic modelling approach? The approach has two generic features (cf. Fig. 1.4).

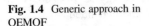

Fig. 1.4 Generic approach in OEMOF

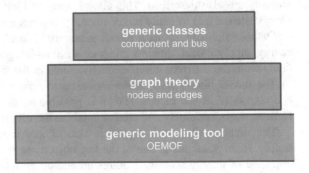

What provides the basis of the generic modeling created by OEMOF? The generic approach is created on two levels: First, the constituents are set up and divided into different sets of constituents. Three types of constituents are defined: These are sources, sinks and transformers. These can represent the following constituents:

- Source: For example, there is a clearly defined energy source for each, wind parks, solar arrays, coal mining and gas imports.
- Sink: An energy sink is an energy or heating consumer, the export of electricity or an electric vehicle.
- Transformers: Transformers can be power stations, or power–to–gas converters, electrolysis plants, geothermic pumps or a reheater. Gas or heat pipes can also be defined as transformers.

Constituents can also be assigned to other constituents. It is up to the modelers to decide what is or is not a constituent of their model.

The other components are also defined by their generic characteristics:

- Transformers: These can have an indefinite number of inputs or outputs. This means that there can either be one defined input and one defined output, or more than one output, for example in the case of a CHP plant, which produces two different types of energy, heat and power.
A wood burning plant has only one input, the bus described by the term 'wood fuel'. But it has two outputs, one is the energy fed back to the bus electricity grid, the other output flows into the heat pipes. Busses are described hereafter. The inputs and outputs can be computed mathematically and linked by a mathematical equation through conversion factors, which can be calculated

based on parameters such as the heating value of wood and the type of power plant.

Transformers can also be used to model a transmission line, such as a power grid and to represent losses incurred in the process of transmission.

- Sink: A sink has an input but no output.

  The sink constituent can be used to model an end consumer, such as a one-family household.

- Source: A source has no input, but precisely one output.

  A constituent of the type source can be used to model wind power plants, or solar arrays or coal plants, or the import of petrol or natural gas.

There are also some constituents listed on oemof.solph that are not generic, like for example:

- General storage system, e.g., batteries or compressed air storage systems.
- CHP plant with extraction condensing turbine using a linear approach.
- CHP plant with extraction or back-pressure condensing turbine (mixed-integer linear approach).

One additional generic element is the so-called bus. A bus corresponds to a node or a cell in the energy system. A bus does not represent a physical component, such as a wind power plant or the grid, but more a virtual element that is required for calculating costs. A bus has the following characteristics:

- Bus: A bus is composed of several inputs or outputs.

  A bus corresponds to a node or a cell in the energy system that must be balanced in every time step. This condition means that the sum of all inputs and outputs at any point in time must be zero.

The second level on which this generic mapping takes place is the graph theory governing the level described in Fig. 1.4. A model in OEMOF is represented by edges and nodes. These edges and nodes are also abstract objects that only receive their specific significance through the parameters that are set by the modeler. More information about graph theory will be given in Chap. 3.

A node can either be a bus or a component. That is why a node is a very versatile element. It can either map a specific region, or one part of the grid of one region, or it can represent different temperatures in a distance heating pipeline system. The modelers determine what each node represents by their setting of parameters.

The edges map the energy flow within the model. Arrows describe the direction of the energy flow.

By means of this abstract mapping of an energy supply system OEMOF provides an approach that enables modelers to represent their specific energy system by means of the generic constituents provided.

What is the process for modeling an energy supply system? The point of departure is the energy support system described in Fig. 1.5. The symbols in Fig. 1.5 have specific meanings that are further described in Fig. 1.7.

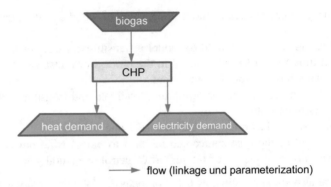

Fig. 1.5 Example of a power supply system

Each individual constituent has its own parameters. For instance the CHP plant has a certain efficiency rating, biogas may generate variable costs and may have a limited quantity. The required heat and electricity may depend on time series (load profile) (cf. Fig. 1.6).

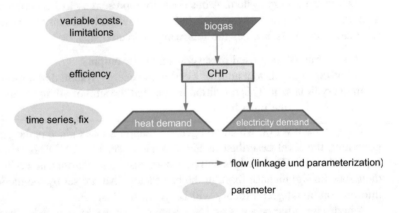

Fig. 1.6 Parameters of the individual components of the system

Modelers develop their optimization model based on the structure of the energy system they want to analyze. The modeling itself takes place in oemof.solph. This is where components are linked via edges and buses (cf. Fig. 1.7). A component can be connected to several buses and a bus can be connected to several components. Between them are the edges, which represent energy inputs or outputs of the components. The edges are called flow in OEMOF. The edges, i.e., the flows, are very important, as it is via the edges that an energy flow between two nodes can be assessed, for example by measuring costs. This is shown as a general example in Fig. 1.7.

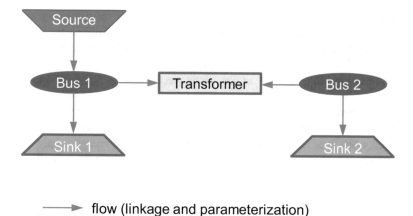

flow (linkage and parameterization)

Fig. 1.7 Generic approach in the program system oemof (arrows = edges) (based on oemof o.J.-a)

This approach is based in graph theory and is also applied for Petri networks. The graphs used for modeling are so-called bipartite graphs. A bipartite graph is a mathematical model that represents the relations between elements of two sets. A special feature of bipartite graphs is furthermore that there is no direct relationship between the two sets, i.e., no intersection. In OEMOF one of these sets are the components, the other set are the buses. Components cannot be directly linked to buses.

By means of these modeling tools, OEMOF provides a linear or mixed-integer optimization model, which is also the case for GAMS (General Algebraic Modeling System) as well as in some other modeling systems. As is the case for other models, the completed model can then be passed on to a solver to solve the equations created as part of the model.

The OEMOF programming system is written in the programming language Python. This is an open source language which also runs on all operating systems. Other well-known programs, such as Google and YouTube, are also based on Python (Python o.J.-b). This language is easy to learn, as it is very simple and clear. Python is a higher, so-called interpreted programming language. It is very versatile and can be used for software development. The programming code is characterized by a clear and easily readable syntax.

Python is an interpreted language, as a software program, the so-called interpreter, reads and analyzes the source code while the program runs and executes the commands via a platform (XOVI o.J.; bib o.J.).

Higher programming languages differ from simpler programming languages based on the level of abstraction that can be reached in the program. Higher programs allow complex logical relations to be written. These can then be read and executed by means of the interpreter.

As OEMOF was conceived based on freely available programming languages, this system of programs is not only fully transparent (Open Source) but can also be applied and modified by the modeler without restrictions. This means that modelers can make changes and adaptations and can then make their modifications available to the OEMOF community. The software can be downloaded online free of charge.

As OEMOF is based on a generic approach and can be adapted to the users' requirements, it is very versatile in its application.

This versatility also extends to the many very different types of optimization functions that can be carried out in OEMOF. OEMOF allows the modeling of linear problems, such as economic supply functions by means of an LP, Linear Programming approach, as well as mixed-integer linear programming (MILP), which could be used as a strategy blueprint for the efficient operation of a complex energy supply network. The following chapter will describe LP and MILP models in a little more detail.

1.3   Linear Programming (LP) and Mixed-Integer Linear Programming (MILP)

Many problems that require answering in business, the economy and the administration and even in our everyday lives are optimization problems. These are often planning or decision-making questions. Examples for problems that are based on an optimization approach are:

- Deciding a location for a school, a manufacturing base, a warehouse. Factors that might influence the decision include distance, connections to public transport or motorways, number of inhabitants, etc.
- Planning a route for a tourism company or a delivery company. This might mean finding a route that can access defined key destinations in the shortest possible time.
- Warehouse planning for incoming and outgoing parcels for companies such as the postal service or Amazon. Parcels are required to be stored and retrieved either by warehouse workers or warehouse robots using the shortest possible route and in the shortest possible time.
- Profit maximization for a company. In production management, this involves decisions such as which product should we produce, made from which raw materials, in what overall time and in what order of processing steps and what price.
- Optimizing aerodynamics, e.g., of a car.
- Optimizing air travel routes to reduce noise level in residential areas.
- Planning and designing an energy supply system for a municipality. Questions that must be answered include which technology is chosen, making use of which energy source and providing which energy output. Relevant factors here include investment costs and energy yield.

Optimization models are particularly useful for network planning and energy supply reliability. Typical problems to be solved in such a context include the type of power plant that will be able to yield the required energy at a specific point in time while offering the lowest possible energy prices. Another relevant factor for planners is the requirement to generate the lowest possible emissions and to use as little energy resources as possible. Another problem might include achieving this with a limited manufacturing capacity. An optimization model might enable planners to determine the best energy supply option. Thus the model is not simply producing an extreme value as its output, but the best or worst value under specific conditions. If the model includes a sensitivity analysis, several alternative options can be generated. Sensitivity analyses are created by varying some key parameters, such as the price of fuel, such as oil or gas. Here, different scenarios can be played out, e.g., one scenario, where the price of gas or oil increases by 10% within the next 5 years and one scenario where the price increases by 20%. Another such analysis might be looking at different possible CO_2 emission limits.

Coming up with a system for solving an optimization question can be difficult. First question to be answered here is whether this problem can be described by integer variables. This would be the case if the question was how many power plants of a certain type are to be built at a certain location. As the objective function itself as well as some of the constraints might contain some nonlinear terms, this must be taken into account when looking for the right model to cover these aspects. For example, the electricity requirement of a private household is not constant across the day but experiences a lot of fluctuation. This will require to be represented by nonlinear equations. But wherever possible, it should be attempted to describe these by means of linear or integer functions. In order to be able to classify optimization problems, the following table lists specific problem classes (cf. Table 1.1). The problem categories are based on a mathematical description of different optimization problems. If purely linear equations can be formulated, this is defined as a linear optimization. An optimization problem is also called a program both in mathematics and in computer science.

Problem class	Acronym
Linear programming	LP
Integer programming	IP
Mixed-integer linear programming	MILP
Nonlinear programming	NLP
Mixed-integer nonlinear programming	MINLP
Global optimization	–

Table 1.1 Classification of optimization problems into classes of problems

Each problem class requires specific algorithms for solving the underlying system of equations. The solving algorithm is stored in the so-called solver as a separate mathematical program. For linear and mixed-integer linear optimization problems good solvers are already available to the modelers. Specific modeling systems, such as GAMS or OEMOF already provide a certain number of solvers to their modelers ready for immediate application. In some cases, licenses must be obtained, other solvers do not require a license.

Linear relations are always easier to map than nonlinear relations. Also, linear equations normally have a global optimum as a solution whereas nonlinear equations might have several different local maximum levels. If it is at all possible to reduce a specific problem to a linear equations system, this is always the preferable option. If a problem cannot be described by a linear system of equations, it is recommended to try and describe aspects of the problem by integer relations, for example the electricity consumption profile of a household. For these types of problems, various good solvers are available. For this reason, Linear Programming and Mixed-Integer Linear Programming are introduced in more detail in the next section.

How does model creation work? Figure 1.8 describes how a problem can be modeled.

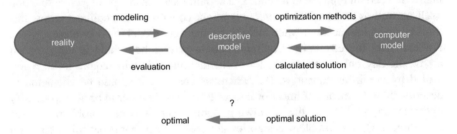

Fig. 1.8 Formal approach to the development of a computerized optimization model (based on Suhl 2006)

The modeling process can be subdivided into seven steps (Domschke 2011):

1. The first step is recognizing and analyzing the problem. One example might be the option for an energy supply company to offer new products to their clients or to expand the types of energy sources or to replace existing power plants by newer and more modern plants. Problems arising from this opportunity are the decision where to put these new plants as well as the requirement to determine the required energy yield. The new opportunity requires the energy company to act and to take a decision.

2. The next step is to define the options and to formulate the objectives. In order to be able to take the right decision, the energy supplier must first decide which objectives it wants to achieve. Is the most important objective to achieve the expansion for the lowest cost or is it more important to minimize CO_2 emissions? There are also often several alternative options for achieving the same objectives, that must be determined and then weighed up against each other. Since not all aspects for all options can be described in detail, e.g., due to missing or inaccurate data or because of time or budget limitations, a simplified representation of the situation will be created. This is the so-called descriptive model. The descriptive model must formulate the optimization problem as

clearly as possible, in order to determine what is to be decided. This will then be formulated as a mathematical objective function for which either the maximum or the minimum level is to be found. The different decisions or the different options are the variables in this mathematical model. Variables are normally represented by a vector $x \in R^n$. When solving an optimization function, a value for each variable of the vector x is determined.

3. The next step is to look at the constraints that limit the system. One such limit or one such constraint could be an upper limit for CO_2 emissions that must not be exceeded. It is important to distinguish between an upper or lower limit or an objective value, which is a value that should be as low or as high as possible but is not a constraint. These conditions are then formulated as constraints in the mathematical model. These constraints are formulated as equations or inequations. If a solution is computed for the vector x the computed values must also be tested against the set constraints. If the constraints are fulfilled for solution x, then x is called a valid solution.

4. The mathematical model will be based on the descriptive model. The variables that are to be calculated have been chosen and the contraints have been set.

5. The next step is data procurement, which can be difficult, either because data are not available in the required format or because they are not available for the required period. This step sometimes requires further models and methods for the gathering of prognostic data.

6. To solve the mathematical optimization model an algorithm (an instruction for solving the problem) is required. By means of the algorithm the data input is transformed into data output step by step. If the algorithm is carried out by a software program, this is called a solver. A solver is a special type of software program designed to solve mathematical problems arithmetically. The solver computes the best or optimal solution based on an objective function.

7. In the next step the computed solution must be assessed. This is normally done by means of a filter (b-wise GmbH o.J.). One such filter could be: Is this solution technically viable, can the required network be built? Are any laws, contracts or legal requirements violated by this solution? Is this a fair solution that is also compatible with our self-image? By means of these assessment criteria the solution can be evaluated as being either viable, requiring modification or being not viable. Depending on the result of the assessment, the model might have to be modified accordingly and a new solution has to be sought.

Once these seven steps have been carried out and the model development process has been completed, the result is a mathematical model capable of computing valid solutions. Methods for formulating the mathematical model are Linear Programming and Mixed-integer Linear Programming. Which of these two methods is used depends on the problem that is to be solved as well as on the shape and variables involved in the mathematical module.

There are different videos available on youtube introducing the topic optimization models. One such video tutorial refers specifically to the modeling of energy supply systems. Some of these videos are available in German, others are available in English (https://www.youtube.com/watch?v=19-XD2E_obg&t=2s).

The videos discuss the questions:

1. Why is it so difficult to optimize an energy supply system?
2. What does creating a model involve?
3. What is meant by optimization?
4. How can I transform a mathematical optimization model into a computer program?
5. What programming language is available for doing this?
6. How can I solve my optimization function?

The videos on this site give an insight into energy system modeling and into mathematical approaches towards formulating optimization problems in mathematical terminology, such as Linear Programming (LP) and Mixed-integer Linear Programming (MILP).

1.3.1   Linear Programming (LP)

One category of optimization models are the so-called Linear Programming Models (LP models). These are often called bottom-up models, as they contain very detailed information about technology and costs (Claussen 2001). Consumer consumption is always calculated on the basis of a load profile. But in some cases consumer behavior can also be integrated as reflection of reality as a non-quantitative factor into the model. However, in order to use such non-quantitative factors, other types of models, such as probability models would have to be used to simulate consumer consumption.

Linear optimization is a subdiscipline of mathematical optimization, which itself is part of operations research. Operations research is a discipline that is concerned with operationalizing different methods from mathematics, statistics, probability, and game theory for solving decision-making problems by research methodology.

Linear Programming as a method consists of formulating a linear equations system and is a mathematical optimization method which computes the maximum or minimum value for a linear function, the so-called objective function. This objective function has n degrees of freedom and is limited by various constraints, which are also formulated as equations or inequations. These constraints act as limits to the system of equations. Constraints could include limited resources formulated as maximum loads or maximum yields or maximum storage capacity or legal requirements to limit emissions to a maximum.

One well-known problem often formulated as an optimization function is the Travelling Salesman Problem (TSP). The TSP consists of a logistical requirement to visit a predefined number of destinations using the shortest possible route. Figure 1.9 presents such a problem example as well as its solution. The defined task

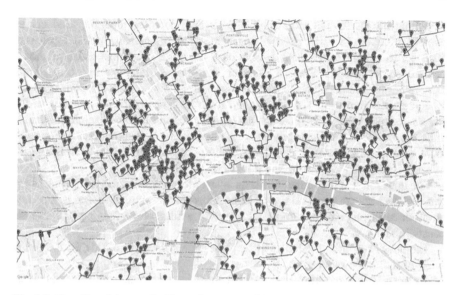

Fig. 1.9 Travelling Salesman Problem: shortest possible walking tour including all pubs of the United Kingdom, London example (HIM o.J.)

is: 'Find the shortest route to visit some 24,727 stops found on the website Pubs Galore—The UK Pub Guide' (HIM o.J.). So the research team had to find the shortest possible walking tour including all pubs of the United Kingdom. To solve the problem they used the TSP cutting-plane method.

Each destination is to be visited only once. The degree of complexity of a TSP problem depends on the numbers of destinations that must be visited. As early as 1954 solution approaches were developed that were based on linear programming (HIM o.J.; TSP 2015).

1.3.2 Example Biogas Production

Before introducing the formal structure of the equations in a linear program, we will first have a look at a specific example, an optimization problem from the area of biogas production.

An agricultural entrepreneur runs two biogas test plants at two different locations, test plant A and biogas test plant B. The energy sources for both of these are liquid manure and horse droppings. The biogas produced in test plant A is used for electricity and heat generation in a CHP plant. Test plant B produces biogas which is then further processed into methane to be used as vehicle fuel. In order to reduce the potential complexity of the equation, we will leave out the calculation as to how much biogas is required to produce one liter of methane. As production takes place at separate facilities, this does not have to enter the equation.

Selling biogas for the production of heat and electricity nets the producer an overall price of 75 € per m³ of biogas and for biomethane a fixed price of 50 € per m³ biomethane. Plant A runs on 8 m³ horse droppings and 2 m³ liquid manure per m³ biogas. Plant B runs on 3 m³ horse droppings and 6 m³ liquid manure per m³ biomethane. Both the biogas and the biomethane are stored in a gas storage unit before sale. The gas storage for biogas has twice the capacity of the storage for biomethane. This means that the production of biomethane from plant B must not exceed half the capacity of biogas from plant A. Another factor that needs to be taken into account is that the overall energy source available is 126 kg of horse droppings and 84 kg of liquid manure.

The question that the farmer needs an answer to is: Which amount of each of the two fuels/energy sources (plant A: biogas and plant B: biomethane, in m³) should he produce in order to maximize his turnover while still complying with the set constraints?

The answer to this question will help the entrepreneur to decide on the future setup of these and other plants.

To answer this question the problem must be formulated as a mathematical equation, which can then be solved in four steps:

Step 1: The gathered data are transformed into a matrix (cf. Table 1.2):
Step 2: The next step is to formulate the mathematical equations underlying this problem. (cf. Table 1.3). For this to happen the decision variables must first be determined. These are variables x and y. Referring to the question described above, these are the amounts (m³) of biogas and biomethane respectively.
Step 3: A graphical representation of x, y in a coordinate system.

Table 1.2 Matrix example *Biogas production*

	A	B	stock (kg)
Decision variable	x (biogas [m³])	y (biomethane [m³])	
Horse droppings (kg/m³ gas)	8	3	126
Liquid manure (kg/m³ gas)	2	6	84
Retail price (€/m³ gas)	75	50	

Table 1.3 Mathematical formulation of the problem

	Mathematical formulation
Objective function	$z(x, y) = 75x + 50y \rightarrow$ Max!
Constraints (CS)	$8x + 3y \leq 126$ (CS1)
	$2x + 6y \leq 84$ (CS2)
Further constraint applies: $y \leq 2x$	$2x - y \geq 0$ (CS3)
Nonnegativity constraint	$x \geq 0$ (CS4)
	$y \geq 0$ (CS5)

Mathematical linear equations normally have a form such as (s. Eq. 1.1):

$$y = ax + b \tag{1.1}$$

Factor a represents the gradient of the line and b the axis intercept on the y axis. In order to illustrate this, a is given the sample value 2 and the axis intercept b is assigned the value 10. By means of these values Eq. 1.1 can be substantiated (cf. Eq. 1.2):

$$y = 2x + 10 \tag{1.2}$$

Equation 1.2 is a specific linear equation that can be translated into an x, y coordinate system, (cf. Fig. 1.10).

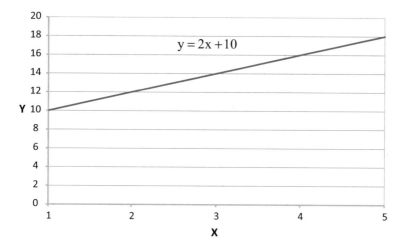

Fig. 1.10 Representation of the gradient and the intercept at the x, y coordinate system

The next step is to review which constraints from our example in Table 1.3 cannot be represented in linear equation (1.2) in their present form. That is the case for CS1 and CS2, which need to be reformulated. Both constraints are solved for y. We will first look at CS1 (s. Eqs. 1.3 and 1.4):

$$3y \leq -8x + 126 \tag{1.3}$$

Solving this for y yields:

$$y \leq -\frac{8}{3} + 42 \tag{1.4}$$

In order to transform these inequations into a Cartesian coordinate system, they must be reinterpreted as equations (cf. Eq. 1.5):

$$y = -\frac{8}{3}x + 42 \tag{1.5}$$

This is a normal linear equation with the gradient $-\frac{8}{3}$ and the axis intercept +42, which can be represented in a Cartesian coordinate system (cf. Fig. 1.11).

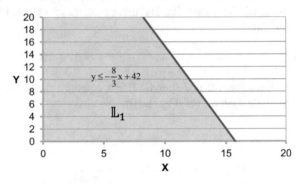

Fig. 1.11 First constraint CS1 as a linear equation in a Cartesian coordinate system

The shaded surface in Fig. 1.11 represents the constraint CS1. Next we will look at the representation of CS2 (cf. Eq. 1.6):

$$2x + 6y \leq 84 \tag{1.6}$$

Again we solve this equation for y and formulate the following linear equation (cf. Eq. 1.7):

$$6y \leq -2x + 84 \tag{1.7}$$

Interpretation as a linear equation (cf. Eq. 1.8):

$$y = -\frac{1}{3}x + 14 \tag{1.8}$$

This equation can again be represented by a graph (cf. Fig. 1.12). All points within the area shaded in green fulfill CS2.

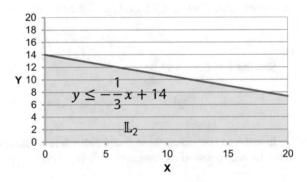

Fig. 1.12 Second constraint CS2 as a linear equation in a Cartesian coordinate system

Now let us have a look at the third inequation CS3. This states that the farmer may produce a maximum of twice the amount of biogas in plant A (2 m³ biogas) as the biomethane produced in B (1 m³ biomethane). The mathematical equation derived from this information is this (cf. Eq. 1.9):

$$y \le 2x \tag{1.9}$$

Formulated as linear equation (cf. Eq. 1.10):

$$y = 2x \tag{1.10}$$

If this equation is represented in a Cartesian coordinate system the result is Fig. 1.13.

Fig. 1.13 Third constraint NS3 in the Cartesian coordinate system

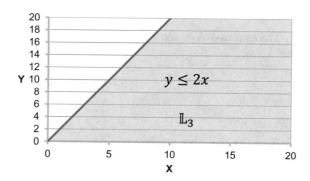

After all the constraints have been solved for y and written as a graph, the surface area and the lines derived can be added together. The solution is the intersection shown in Fig. 1.14.

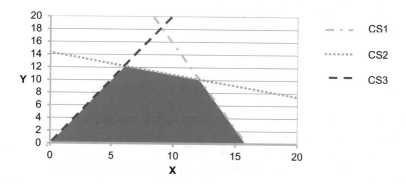

Fig. 1.14 Intersection of the inequalities in the Cartesian coordinate system

The shaded surface between axis and line represents the values for which the constraints are fulfilled. All points within the area shaded in blue are valid results for the inequations system and thus form the solution set. But the optimum must still be determined. The equation can be solved by means of a graph or by means of calculation. The next step shows the solution by means of a graph. The following step is the calculation of the solution.

Step 4-a: Solution by means of a graph

This is the moment to use the objective function, for which a maximum is sought. In order to be able to represent the results in a Cartesian coordinate system, the objective function must be solved for y. Furthermore, the objective function must be written as an equation. To ensure that the equation starts from point zero on the coordinate system, the target value z is set to zero (cf. Eq. 1.11).

$$z(x, y) = 150x + 100y = 0 \qquad (1.11)$$

The next two steps show the transposition towards a solution for y (cf. Eqs. 1.12 and 1.13):

$$100y = -150x \qquad (1.12)$$

$$y = -1.5x \qquad (1.13)$$

Now the objective function can be represented in Fig. 1.14 based on Eq. 1.13 (cf. Fig. 1.15).

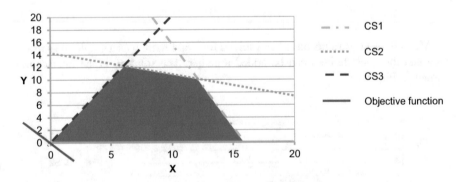

Fig. 1.15 Objective function at point 0 in the Cartesian coordinate system

If the graph of the equation is then moved up or down parallel to the zero-value equation, the maximum or the minimum can be determined. As for this equation, a maximum for z is required, we transpose Eq. 1.13 upwards parallel to the linear equation with the zero value. The line is transposed until it touches the final point of the solution set. This point (marked in red) represents the optimum solution (cf. Fig. 1.16).

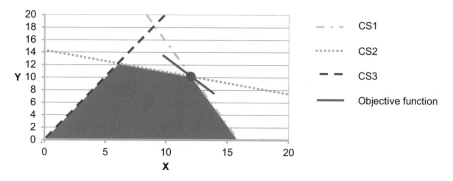

Fig. 1.16 Graphical solution of the linear optimization problem

Now we have provided a graphical solution, the next step is to calculate an optimum.

Step 4-b: To obtain the calculation, it is important that the optimum will be at one corner of the solution set. This is based on the knowledge that the optimum will have to be part of the solution set presented in Fig. 1.14. But there can only be one optimal point. As we have seen in the graphical solution, there are several valid solutions, but not one optimum visible. The x and y values can be read from Fig. 1.14:

$$P(0;0), P(6;12), P(12;10), P(15,75;0)$$

In order to determine which of these points represents the optimum, these values are entered into the objective function (cf. Eq. 1.14):

$$z(x, y) = 150x + 100y \tag{1.14}$$

The following solutions emerge (cf. Eqs. 1.15–1.18):

$$z(0;0) = 150 \cdot 0 + 100 \cdot 0 = 0 \,€ \tag{1.15}$$

$$z(6;12) = 150 \cdot 6 + 100 \cdot 12 = 2.100 \,€ \tag{1.16}$$

$$z(12;10) = 150 \cdot 12 + 100 \cdot 10 = 2.800 \,€ \rightarrow \text{Maximum} \tag{1.17}$$

$$z(15,75;0) = 150 \cdot 15,75 + 100 \cdot 0 = 2.362,50 \,€ \tag{1.18}$$

Hence there is a maximum, which answers the farmers question: Which amount of each of the two fuels/energy sources (plant A: biogas and plant B: biomethane, in m³) should he produce in order to maximize his turnover? He should produce 12×1 m³ biogas in plant A and 10×1 m³ biomethane in plant B.

1.3.3 Formal Structure of LP

The formal structure of a so-called Linear Programming Model (LP Model) for the mathematical formulation of an optimization problem (or a decision task) consists of the following three components (cf. Table 1.4):

Table 1.4 Formal structure of a so-called linear programming model (LP model)

	Mathematical formulation
Objective function The goal of an LP model is to maximize or minimize a function depending on one or more variables. The linear function that is to be optimized is called the objective function. The variables contained in it $(x_1, x_2, \ldots, x_p)$ are called decision variables. The factors c_n are real numbers.	$z = z(x_1, x_2, \ldots, x_p) = c_1 x_1 + c_2 x_2 + \cdots + c_n x_n \to \text{Max}!$
Constraints (restrictions, limits) The factors a_{pn} to b_p are also real numbers.	$a_{11}x_1 + a_{12}x_2 + \cdots + a_{1n}x_n \leq b_1$ $a_{21}x_1 + a_{22}x_2 + \cdots + a_{2n}x_n \leq b_2$ ... ... $a_{p1}x_1 + a_{p2}x_2 + \cdots + a_{pn}x_n \leq b_p$
Nonnegativity condition In many decision-making problems, the variables must not take on negative values. Looking at our example, it is not possible to produce negative quantities of biogas. For this reason, it makes sense to restrict the result value for x to be greater or equal to zero.	$x_1 \geq 0$ $x_2 \geq 0$ ... ... $x_n \geq 0$

Once the equations or inequations have been created, the next step is solving the equation or solving the system of equations. A simple graphical or an arithmetic solution are normally not enough for a complex problem within the energy sector, as these problems normally require the most efficient way of dealing with something, including using the least amount of computer power possible, which is normally the case when using LP modeling. The search for the best solution is conducted by means of algorithms. There are some standard algorithms available for solving LP models, such as the simplex algorithm, which is the algorithm most commonly used at the moment. The simplex algorithm method was created and introduced by Dantzig in 1947 (Domschke 2011). As the simplex algorithm is very sophisticated, its methodology is not described in detail here. The basic methodology of the simplex algorithm is to study the margins of the solution set, as it is already clear that the optimum will be found at the bottom or top margin of the valid solution area. These marginal values are then included in the objective function in order to find the solution. The algorithm will then repeat this process time and again, jumping to the next extreme. The principle guiding this progression along extreme values is that the next value included in the objective function must not be lower than the previous one. Once a next, higher marginal value has been found, the iteration continues to the next corner. If none of the adjoining corners is higher, the values found must at least be a local maximum. If there is no other solution equal in value, then this value must be the optimum and thus the solution to the optimization problem. The simplex algorithm is one of the most powerful tools for solving LP problems (Domschke 2011).

1.3.4 Mixed-Integer Linear Programming (MILP)

The difference between linear programming and mixed-integer linear programming mainly lies in adding more restrictions. One key constraint for MILP is that some variables must have integer values. Particularly relevant are binary, i.e., 1 or 0 variables. These allow for a very simple mathematical formulation for a decision. For example, this variable type could be used to decide if a wind power plant with a certain yield should or should not be built at a specific location. If the binary variable takes on the value 0, the recommended decision is not to build the wind power plant. If the variable is returned as 1, this corresponds to the recommendation that the wind power plant should be built. However, the other variables in the model can still take on real number values. MILP problems are a key tool for analysts and engineers making decisions for large projects. Such questions are relevant in the field of location planning but also operation strategies for power plants, in particular where the option is binary (switched on/switched off). There are a number of contexts where integers rather than real numbers are a relevant decision-making option, as in some cases the question might not simply be binary, such as 'Should we build a wind power plant at this location?' but could involve decisions such as 'How many wind turbines should be placed at this location?'. And of course, the number of wind turbines will always be an integer.

We will therefore first look at an optimization problem with integer variables (IP Model). This decision-making problem concerns investment. For one specific region an energy supply system is to be built that covers the energy requirements of its citizens. An investor has been found for this project. The investor is willing to invest a considerable amount of money, in order to provide the best possible energy supply with the current technological possibilities. For this reason, the maximum sought in this case are the highest possible costs. In order to guarantee optimal energy supply to the inhabitants of the specified region, x energy transformers of type1 and y of type 2 are to be bought. Plants of type 1 cost 2 Mio. € and those of type 2 cost 4 Mio. €. This information is integrated into the objective function. Both plants use electricity and process steam. The plant of type A requires 2 t of process steam and plant of type 2 requires 6 t. The storage capacity of the energy storage is limited to 14 t. This formulates our first constraint. Each plant of type A1 consumes 6 kWh electricity and plants of type 2 consume 4 kWh. But together they must not access more than 20 kWh electricity, which is our second constraint. As plants can only be counted in integers, variable x and y both must fulfill the integer and the nonnegativity constraint. What is the maximum of plants that can be bought?

The objective function will look like this (cf. Eq. 1.19):

$$z(x, y) = 2x + 4y \rightarrow \text{Max!} \tag{1.19}$$

Including the constraints (cf. Eqs. 1.20–1.22):

$$2x + 6y \leq 14 \tag{1.20}$$

$$6x + 4y \leq 20 \tag{1.21}$$

$$x, y \geq 0 \text{ and must be integers} \tag{1.22}$$

This problem can be represented as a graph as in Fig. 1.17.

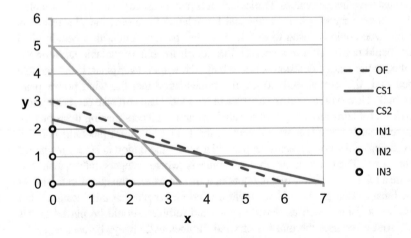

Fig. 1.17 Integer optimization problem, IN: integer number, CS: constraint, OF: objective function

All valid solutions to this function, which also comply with the constraints, are marked as white dots in Fig. 1.17. In the previous biogas example, the solution set was to be found in the intersection of the two linear functions for the constraints. In this IP example the intersection between the lines for Eqs. 1.20 and 1.21 does not offer a valid solution, as the integer constraint is not fulfilled. The optimum is again found by moving the objective function upwards or downwards parallel to its zero value (OF). The highest integral point is $x = (1; 2)$ producing the objective value $z(x^*) = 10$. This result is shown in Fig. 1.18. The highest integral point is marked in green (see the arrow).

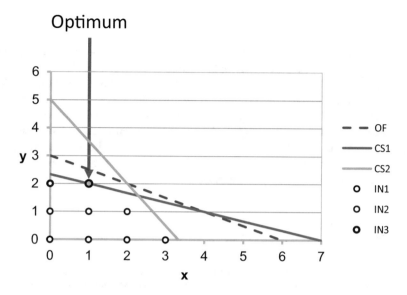

Fig. 1.18 Result of the integer optimization problem, IN: integer number, CS: constraint, OF: objective function

The questioning for a mixed-integer problem would be more complex. One problem might be the decision if type 1 or type 2 should be invested in ($x = 1$ and $y = 1$ resp.) or not ($x = 0$ and $y = 0$ resp.). Other variables might be added to this (such as. x_3, x_4) which can have real number values. A constraint could be added with regard to emissions, e.g., each plant must only emit a certain amount of CO_2. The CO_2 amount could be represented by variable x_3. A further constraint could set a maximal runtime per day for each plant. This time limit could be represented by variable x_4.

1.3.5 Sensitivity Analyses

The parameters inserted in an optimization model might change across time. For example, costs for a wind power plant, electricity consumption for households, the price of energy sources, i.e., silage for biogas plants, but also retail prices for electricity per kWh, etc., all might change across time. This is also true for political regulations for emission limits. Also, some data gathered are not fully reliable, as they might be estimated or prognostic data.

In a sensitivity analysis, the parameters likely to vary in future are researched, in order to study the effect of changes on the optimal solution. This means that the solution to an optimization problem is tested as to the effect a change in input data in the future might have on the current system. This affects both the coefficients of the objective function c_n as well as the right side of the equation b_p and the coefficients of the constraints a_{pn}.

If more than one parameter is varied at the same time, this is called a parametric sensitivity analysis or a parametric optimization (Domschke 2011).

Sensitivity analyses give an important insight into how a decision-making process would be affected by a change in one parameter, e.g., if the energy source price increases or falls sharply. This analysis helps in the decision-making process.

Particularly in the energy sector, where each decision tends to have long-term consequences (such as the lifetime of power plants) and where there are various unavoidable uncertainties (such as oil price, energy demand, supply capacity) reliable analyses are of increased relevance.

References

bib Bildungszentrum für informationsverarbeitende Berufe gGmbH: Programmiersprachen. Paderborn, o.J. https://www.bg.bib.de/portale/bes/Progentwicklung/Programmiersprachen/ Programmiersprachen-110.htm. Accessed on 03 Oct 2017

b-wise GmbH: Problemlösungsmethoden : Lösungen bewerten und entscheiden. Karlsruhe, o. J. https://www.business-wissen.de/hb/loesungen-bewerten-und-entscheiden/. Accessed on 04 Nov 2017

Claussen, E. et al. (Ed.): Climate change : science, strategies, & solutions. Leiden et.al., Brill Verlag, 2001, https://books.google.de/books?id=g85OJ76ufJoC&pg=PA157&lpg=PA157&dq=Linear+Programming +button+up&source=bl&ots=qAbgY4MjcF&sig=CVNMEoG__AsXBO1Y1kJWfLaM-jk&hl= de&sa=X&ved=0ahUKEwjnvIecsNbWAhUEZVAKHUjKAZ0Q6AEITjAJ#v=onepage&q= Linear20Programming%20button%20up&f=falseDatei:LinearProgramming.doc   Accessed on 06 June 2018

Domschke, W. et al.: Einführung in Operations Research. 8. Auflage, Springer Verlag, Heidelberg et al., 2011

Forum für Energiemodelle und Energiewirtschaftliche Systemanalysen in Deutschland (ed.): Energiemodelle zum Klimaschutz in Deutschland : Strukturelle und gesamtwirtschaftliche Auswirkungen aus nationaler Perspektive. Springer Verlag, Berlin (1999)

HIM Hausdorff Research Institute for Mathematics: UK24727 : A shortest-possible walking tour through the pubs of the United Kingdom. Bonn, o.J. http://www.math.uwaterloo.ca/tsp/pubs/ index.html. Accessed on 04 Nov 2017

Kallrath, J.: Gemischt-ganzzahlige Optimierung: Modellierung in der Praxis, 2nd edn. Springer Spektrum, Wiesbaden (2013)

König, C. (ed.): Energiemodelle für die Bundesrepublik Deutschland. Springer, Basel (1977)

Krutzler, T. et al., Umweltbundesamt GmbH (Ed.): Szenario erneuerbare Energien 2030 und 2050. REP–0576, Umweltbundesamt GmbH, Wien, 2016. http://www.umweltbundesamt.at/ fileadmin/site/publikationen/REP0576.pdf, Accessed on 04 Oct 2017

Nagel, J.: Ein analytisches Prozesssystemmodell zur Bestimmung von Rahmenbedingungen für den wirtschaftlichen Einsatz biogener Energieträger im ländlichen Raum, dargestellt an einem Beispiel aus dem Bundesland Brandenburg. Diss., Fortschritt-Berichte VDI, Reihe 6, Nr. 403, VDI-Verlag GmbH, Düsseldorf (1998)

Nagel, J.: Energie- und Ressourceninnovation : Wegweiser zur Gestaltung der Energiewende. Hanser Verlag, München (2017)

oemof-developer-group: Using oemof. Berlin, o.J.-a. http://oemof.readthedocs.io/en/stable/using_ oemof.html#feedinlib, aufgerufen 02 Sept 2017

Pfaffenberg, W., et al. (ed.): Energiemodelle zum europäischen Klimaschutz : Der Beitrag der deutschen Energiewirtschaft. Umwelt- und Ressourcenökonomie, Band 22, LIT Verlag, Münster (2004)

Python Software Foundation: Quotes about Python. Wilmington, USA, o.J.-b. https://www.python.org/about/quotes/. Accessed on 03 Oct 2017

Schultz, R. et. al. (Ed.): Innovative Modellierung und Optimierung von Energiesystemen. LIT Verlag, 2008

Suhl, L. et al.: Optimierungssysteme. Springer Verlag, Berlin et al., 2006

TSP: Queen of College Tours: http://www.math.uwaterloo.ca/tsp/college/index.html. Accessed on 04 Nov 2017 (2015)

Walbeck, M., et al.: Energie und Umwelt als Optimierungsaufgabe : Das MARNES-Modell. Springer-Verlag, Berlin (1988)

XOVI GmbH: Interpreter. Köln, o.J. https://www.xovi.de/wiki/Interpreter. Accessed on 03 Oct 2017

Chapter 2
The Generic Base Model in OEMOF

If you want to create and compute energy supply models using the OEMOF Program, you must have a basic understanding of the Python programming language as well as an understanding of the OEMOF infrastructure. In addition to the many OEMOF specialized libraries, packages, and modules, OEMOF also offers libraries, local modules, and packages in Python. These Python modules and packages are specific program modules for research applications, such as mathematical optimization equations, network analysis, and data analysis.

Datasets are key ingredients for simulation and optimization programs. In many cases, a huge amount of data must be processed for one calculation. In order to have all relevant data stored in one place, they are normally compiled into a dataset. Data that are stored in other files must be included in the program via a specific data interface. One such interface is a CSV file. Another option is to use a SQL database. If GIS data is used, a geodatabase may be a useful format. A geodatabase can be created with a PostgreSQL system based on a PostGIS program. The PostgreSQL system is an object-relational open source data management system. The PostGIS system was created as an extension for the object-relational database PostgreSQL. PostGIS contains geographical objects and functions. Both systems can be integrated into Python applications (oemof 2014b).

Modelers planning to work in OEMOF should note that all texts of the OEMOF website and the examples presented in this book are written in plaintext markup syntax reStructuredText (reST). This markup syntax is used to prepare data for computer processing (Maier 2015). ReST makes the source code accessible to the user. The syntax of reST can best be illustrated by the following example:

attr: 'om.NonConvexFlow.status':

The word 'attr' means 'attribute'. This is followed by 'om.NonConvexFlow. status'. The actual attribute is called 'status'. This status is of the class 'NonConvexFlow'. This expression represents the path where the attribute is found.

© Springer International Publishing AG, part of Springer Nature 2019
J. Nagel, *Optimization of Energy Supply Systems*, Lecture Notes in Energy 69,
https://doi.org/10.1007/978-3-319-96355-6_2

2.1 Python Basics

The OEMOF system of programs offers a framework and a toolbox for the modeling of energy supply systems. This framework, including the toolbox, has been implemented as an object-oriented application in Python. Object-oriented approaches are based on classes and objects. Objects are also called instances. Object-oriented approaches were developed in order to ensure that data and functions that access or compute data are encapsulated in so-called 'classes' to protect them from external access. This means that users of these classes as well as methods and functions of external objects cannot access the objects or change the data itself (Klein o.J.-a).

If you are modeling a program system in order to optimize an energy supply system, you will become aware of how fast the programming code grows. Keeping the code clear and free of errors becomes a challenge. This challenge can only be met if your source code is well-organized. One system for organizing and structuring source code is modularization. Python offers two different types of modules (Klein o.J.-b):

– Libraries
  Python offers general functionalities to all its users. Such functionalities include 'sayhello' but it also defines different data types such as integers, floating point numbers or strings. These are implemented by means of so-called libraries. These libraries include:

  – A comprehensive standard library
  – Libraries programmed by the users
  – Libraries from third parties.

– Local modules
  These are program sections that are only available for one specific program. For OEMOF applications, the modules that have been implemented in Python are of particular relevance. These are made available as separate programs. They all have the file ending '.py'. Local modules are normally used to describe classes and functions of one specific program.

As a rule, modules are integrated into your source code by using an import command. In order to maintain a clear structure, it is recommended that the import command is given at the very beginning of the source code. Several modules can be executed at the same time, separated by commas. A module is imported by the following command:

```
# To import an entire module:

import <modulename>
```

```
# If only one class from a module is to be imported:
```

```
from <modulename> import <classname>
```

By means of the source code, modules are activated and integrated into the user's program.

If several modules are interconnected and there is a file with the name '_init_.py' these modules can be combined into one directory (Klein o.J.-c; Python 2018c). These module combinations are called packages. Several other python modules can be listed in addition to the '_init_.py' file. The file '_init_.py' can either be empty or may contain a code to be executed during the import. A Python package references the directory with the Python modules. Normally Python packages are stored in the following directories:

- For Linux users: /usr/lib/python/site-packages
- For Windows users C:/Python27/Lib/site-packages/

To be able to use packages as part of the users' source code, this package must be initialized.

```
mypackage/__init__.pymypackage/mymodule.py
```

After initialization the package can be imported. This can be done in two different ways. Either a whole module is imported from a package, or only one class of the module can be imported.

```
# Import of a module from a package:
import mypackage.mymodule
```

```
# Import of a specific class from a module:
from mypackage.mymodule import myclass
```

In addition, there is the possibility of renaming a module during import, in order to integrate it into your user code or to reduce the length or the module name:

```
import <modulename> as <freename>
```

2.1.1   Classes, Objects and Methods

First, *objects* must be differentiated from classes. Objects are used to encapsulate data and functionalities. They are instantiated during the running of a program. They are instances of a particular *class*. Classes could also be called 'types of objects'. For example, we can define the *class windpowerplant*. This *class* would then further be defined as containing rotor blades, a rotor hub, a nacelle, a yaw drive, a generator, and a tower. A specific wind power turbine of type A, with specific qualities, such as a rotor blade diameter of 100 m and a rotor hub height of 210 m would then be an *object*. The classes thus define the blueprint of a wind power plant, and the *objects* are specific wind turbines (Wind Turbines of type A). The class describes the structures and behavior that is common to all specific objects. For this reason, objects are often called instances of a class.

Objects have a state, a behavior and an identity. The state of an object is characterized by its qualities (attributes) as well as by its connections to other objects. For example, for our given example of a wind power plant, the rotor speed would be an *attribute*. This speed can be associated with a maximum or a minimum value or a current value. The attribute 'rotor speed' could be assigned a value, such as, the current rotor speed is 12 rpm.

These methods define the behavior of an object. Normally, the behavior of an object is defined through several methods. One such method for the object *windpowerplant A* could be the starting, another method could be stopping the rotation or measuring the current speed of the rotor blades. One other method would be to access and alter the attribute 'rotor speed'. The method also includes the arithmetic operation 'take current rotor speed and increase by factor five'. The new rotor speed would then represent the current rotor speed. This method could be called 'alter_rotorspeed'. This is an access method, as only the methods defined for this class can access attributes, i.e., the data associated with the attribute, in this case the data of the rotor speed. As this is an object-oriented approach, all object data is encapsulated. Only the method of the class has access to the information how the class and the objects are implemented (Klein o.J.-a).

Functions and operations are associated with classes. Functions are processing steps in a computer program. Functions require an input (argument). Once a processing step has been completed, a *functional value* (also called result value) is derived. One example for a function would be for a computer program to derive the name of the bank from the IBAN (International Bank Account Number). For this function, the argument is the IBAN, and the derived value is the name of the bank. In a computer program, this could be encoded in the following way: Bank = IBAN. IBAN is the name of the function. The source code of this function describes how the name of the bank is derived from the IBAN. Operations are subroutines that do not have an argument and do not generate a value. Operations are a sequence of commands that are executed in sequence by a computer program. If a computer program requires the same sequence of commands multiple times, this sequence is then combined into an operation. The sequence of commands is not listed in the user's source code, instead, the relevant operation is listed. Functions and operations are *subroutines*. Subroutines facilitate the writing of source code, provide a clearer structure and thus make it easier to detect programming errors.

Each object is assigned an *identity*, in order to distinguish it from objects of the same state and the same behavior. The identity is stored by means of a specific object key, an identifier inside the program code. The object identifier for a wind power plant of class A could be: WT-A for a wind power plant of class A, and WT-B for a wind power plant of class B. The relevant class would contain the formal information. The class defines the qualities of an object, i.e., which attributes and which methods it contains (cf. Fig. 2.1).

Fig. 2.1 Modeling the class *windpowerplant* with its attributes and methods (based on Klein o.J.-a)

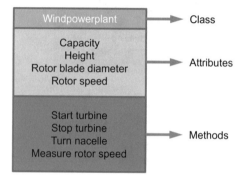

Different classes are often related to each other. Thus the class *windpowerplant* is the parent class, and the other classes are child classes, as there are big and small wind turbines. 'Small wind turbines' and 'big wind turbines' would be children of this parent class. Child classes are derived from the parent class. Thus, specific attributes and methods of the parent class are inherited by the child classes. But child classes also have their own specific attributes and methods (cf. Fig. 2.2).

The attributes are accessed via the methods in order to process or alter them. A method is also a subroutine. In object-relational approaches, methods are associated with classes. In the example here, the rotor speed of the class 'windpower plant' can be altered via the method 'start turbine' followed by the method 'stop turbine'. The method 'measure rotor speed' can be used to derive the current rotor speed. The method would be issued with an argument, for example 5 rpm, which would be the resulting value. A method is implemented once for each object. The syntax for accessing a method is different from the syntax of a function (cf. Sect. 2.1.3).

An object is generated from its class. When the object is created, it is assigned a data type, such as integers or floating decimal point numbers, date or time as well as its attributes and its methods.

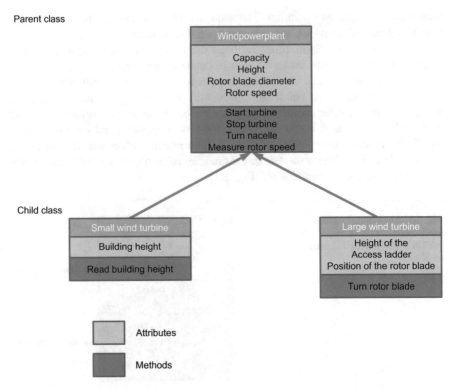

Fig. 2.2 A parent class and its children—the inheritance principle (based on Klein o.J.-a)

2.1.2   Data

The object-oriented approach is based on classes, objects, methods, and data. Data are instances of the attributes of an object. If a method accesses the attribute of an object, it uses this data (e.g., as output, or to alter this data) the attribute itself can be altered. Data is protected from being accessed by other objects by encapsulating data in an object, requiring a method as a means of access to this data. A foreign object would have to both know the object and be in possession of the required access method to access the data. Methods can include algorithms, e.g., for plausibility testing, data type transformations or required calculations (cf. Fig. 2.3).

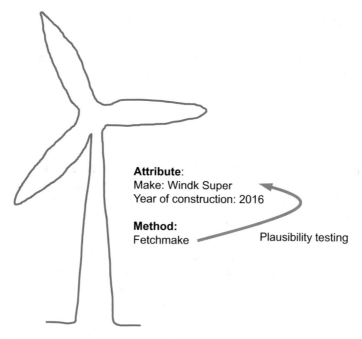

Fig. 2.3 Encapsulation of data (based on Klein o.J.-a)

The attribute *make* (of a wind power plant) would be deposited as a string (of characters). As the data are encapsulated, this information cannot be directly accessed. A method is required to do so, e.g., the method 'fetchmake', which the user can access with his program. The method 'fetchmake' can then provide the name of the *make*. The method 'fetchmake' can also check in which country this make of wind power plant is available and can further check, wether there is availability across the area in which the user operates.

An even better system is to encapsulate the attributes of a class. In order to hide attributes from public access, Python offers the following mechanism as shown in Table 2.1.

Table 2.1 Mechanism for providing accessible and inaccessible attributes (Klein 2018)

Name	Designation	Meaning
name	Public	Attributes that do not start with a lower case hyphen are writable and readable inside and outside their class.
_name	Protected	Can be accessed to write and read from the outside. Developers show in their syntax that these attributes are not to be used.
__name	Private	Attributes starting with a double lower case hyphen: cannot be accessed or seen from the outside.

2.1.3   Creating Classes, Objects and Methods

How are classes created in Python? For this we will use our example, the class *windpowerplant*. This is created in the following way:

```
class windpowerplant:
    pass
```

The class is introduced by the keyword 'class'. So far, the class *windpowerplant* does not yet include attributes or methods. The keyword 'pass' informs the interpreter that the required instructions will follow at a later stage. Thus the syntax for defining a class consists of two parts.

- Head: The head normally consists of the keyword 'class' followed by a space and the name of the class, in this example *windpowerplant*. If the class is a child class, brackets behind the class explain from which parent class it inherits its methods and attributes. For example, if there is a child class *small_wind_turbine* that inherits from the parent class *windpowerplant*, the child class would be written as: 'class small_wind_turbine(windpowerplant)'. If no parent class exists, brackets are not applicable in Python. The final character in the string is a colon.
- Body: Indented in the body, one or several instructions are given. In the example, the instruction starts with the keyword 'pass'.

Class and *Pass* are two of 33 keywords in Python, these are listed in Table 2.2. For the class *windpowerplant*, now two instances *a* and *b* are to be created and attributes assigned.

Table 2.2 (Often used) keywords in Python (ZetCode o.J.)

False	Continue	For	Lambda	Try
None	Def	From	Nonlocal	While
True	Del	Global	Not	With
And	Elif	If	Or	Yield
Assert	Else	Import	Pass	
Break	Except	In	Raise	
Class	Finally	Is	Return	

```
class windpowerplant:
    pass


a = windpowerplant()
b = windpowerplant()


a.make = 'Windk Super'
a.construction_year = 2016
b.make = 'Windk Plus'
```

```
b. construction_year = 2017
print (a.make)
Windk Super
```

Two instances of the class *windpowerplant* are initialized as *object a* and *object b*. Empty brackets behind the instance (*a = windpowerplant()* suggests that the class *windpowerplant* has been instantiated already and that this is an object and not a class.

The make and the date of production are assigned to each object instance. Attributes are added after the name of the instances, separated by a dot. The attributes are called instance variables. These variables are only defined within each object, i.e., the instance variables of *object a* that define this instance. The corresponding term for classes are class variables (Python 2018a). This means that attributes are defined directly for the class. The command *print(a.make)* asks for the output of the make of the wind power plant of instance a. The output in this instance is *Windk Super.*

Methods are defined for each class. They are introduced by the keyword *def.* In Python methods differ formally from functions in two key aspects (Klein o.J.-a):

- Methods are functions that are predefined within a class.
- The first parameter of a method always refers to the object for which it is executed. That is why methods always start with the parameter itself followed by a reference to an instance of a class. In the following example, the method *sayhello* is not executed on its own, but is followed by the accessed instance *a,* separated by a full stop. This means that the command is referred to the instance *a* by the parameter *self.* Object *a* is an instance of the class.

As an example for a method, we will look at a wind power plant's response to the method *sayhello.* If this method is executed, the output is the value *hello.* That may not make a lot of sense, but it illustrates the execution of the method.

```
class windpowerplant:

    def Sayhello(self):
        print('hello')

if __name__ == '__main__':
    a = windpowerplant()
    a.Sayhello()
```

The line of code: *if __name__ =='__main__'* has the following meaning:

- __name__: This is a so-called *built-in attribute.* The name of a module is defined within this built-in attribute. The module here represents the method *print.* This method is stored under the name *print.py.*

– __main__: If this module is to be imported into the programming code, this is done by means of the command *import print* at the start of the programming code. The built-in attribute '__name__' has the value *print*'(__name__ == '__print__'). But it is also possible to execute the method *print*, i.e., the file *print.py* as a separate program. In that case, the program is executed as shown above, the required value is indicated by '__main__'.

The query 'if' gives the following instruction to the program: If the program is run as a main program (__main__) and was not imported, executed the following method.

The syntax 'a.sayhello' requires an output of the value *hello* only from the instance *a*. This is how the method is referred to a specific instance.

Functions have a very different setup structure. The following example shows how the pressure measured is converted from bar to pascal (Pa) by means of a function.

```
def pressure(bar_in_Pa):
    " " " returns the pressure in pascal " " "
    return bar_in_Pa * 10e5

for p in (22.6, 25.8, 27.3, 29.8):
    print(p, ':', pressure(p))
```

The output value is

```
22.6 : 2260000
25.8 : 2580000
27.3 : 2730000
29.8 : 2980000
```

The example shows that functions have the following structure:

```
def function-name(parameter_list):
    instruction(s)
```

That is how a function differs formally from a method.

2.1.4   Class Variables

Let us now have a look at how to create attributes for a class. The instances of a class are not visible to the class. That is why the parameter *self* has to be placed at the beginning. Attributes of an object are defined within a class by means of the syntax *self.attributename = value*. The creation of attributes must take place when an object is initialized. Within a class, the predefined attributes can be accessed via the syntax self.attributename. Outside a class, attributes are accessed via the syntax *instance.attributename.*

In our examples, the variables *make* and *construction year* have not yet been defined within the object and are therefore not yet known. This means that the values of these attributes have to be assigned within their class. Methods are required to define theses values. There is also a specific syntax for executing methods. Methods are always executed within a class via *self.methodname()* and via *instance.methodname()* outside a class.

In the following section, two objects of the class windpowerplant are assigned a make and a construction year via *setmake* and *setconstructionyear.*

```
class windpowerplant:

    def sayhello(self)
        print('hello, I'm of the make' + self.make)

    def setmake(self, make):
        self.make = make

    def setconstructionyear(self, construction_year):
        self.construction_year = construction_year

if __name__ == '__main__':
    a = windpowerplant()
    a.setmake('Windk Super')
    a.setconstructionyear(2016)
    b = windpowerplant()
    b.setmake('Windk Plus')
    b.setconstructionyear(2017)
    a.sayhello()
    b.sayhello()
```

The character '+' in the code indicates that the function *sayhello()* is additionally to generate the string 'hello, I'm of the make of' followed by the relevant value of the parameter *make*. The output for object *a* will look like this:

```
'Hello, I'm of the makeWindk Super.'
```

Another important method that needs to be introduced here is the method *__init__*. This method automatically defines the attributes of an instance after its creation. This method performs the initialization of an instance. In the source code of a class definition, the method *__init__* should be placed directly below the class headers. The example below, the definition of the class *windpowerplant* illustrates this:

```
class windpowerplant:
    # Here a simple class 'windpowerplant' is generated.

    def __init__(self, distance, noises):
        # Initialize a new windpowerplant.
        # Arguments are:
        # distance (string): distance to the next building.
        # noises (vsh- noise when turbine is rotating
        self.distance = distance
        self.noises = noises

    # A method to return the distance and render the noises is provided.

    def show_distance(self):
        print self.distance
    def playnoise(self)
        print self.noises + '!!!'
```

The example above shows the definition of the class *windpowerplant*. The first paragraph introduces a short description of the class indicated by the hashtag (#).

The next line, executing the method *_init_*, defines how instances of the class *windpowerplant* are generated. The method *_init_* is automatically executed whenever the name of the class is run as a function. The arguments that are key attributes of the instances are given in brackets behind the method. The relevant arguments are distance to buildings and noises caused by the rotation of the blades. This also means that the two defined functions defined here *show_distance()* and *playnoise()* are available to the object. These functions allow the output of relevant object variables.

In the following section, a new instance of the class *windpowerplant* is generated.

```
# Initializing of a new instance:
my_windpower_plant = windpowerplant('3 m', 'vsh, vsh, vsh')

# Functions get tested:
my_windpower_plant.show.distance()
# The result is: '3 m'

my_windpower_plant.playnoise()
# The result is: 'vsh, vsh, vsh !!!'
```

2.1.5 Inheritance

Inheritance is a very important principle in object-oriented programming. That means that a general class is defined and that the attributes and methods of this class can be inherited by its subclasses. The subclass can have additional attributes not contained in the parent class and inherited attributes can also be modified by the subclass. In the source code inheritance is signaled by adding the name of the subclass in brackets behind the class definition. The following example shows how a subclass is formed based on the parent class *windpowerplant*.

```
class subclass(windpowerplant):
    pass
```

The examples of code and commands presented here are to help readers understand key aspects of programming in Python. For more information, please follow the references listed below. Links to relevant references can be found here: https://python-verband.org/Members/reschke.michael-40gmail.com/ressourcen. The references listed here are organized according to topics, such as Introduction, Big Data or Databases. This helps to select relevant literature.

2.1.6 Programming Your Own Analyses in Python

The Energy Turnaround affects vehicle mobility. Different types of fuels are available, such as diesel fuel, ethanol, or biogas. A big debate has begun whether we should move to electric vehicles instead. The decision which mobility concept is to be preferred and which energy source will be a viable option for the future can be supported by using an optimization model.

The following example will be used to show how Python can be used for programming an optimization model. To set this up, the class *car* is introduced as parent class and two child classes, *burners* and *evehicle*. Two methods are defined for the parent class *car*. For the child classes additional methods are created, that apply only to those classes. Furthermore, additional parameters are created, such as consumption cost or overall cost. These parameters are assigned values, which will be redefined, i.e., overwritten once calculations have been performed.

This example can be taken as a model to take your first steps using Python as a programming language. The child class *evehicle* is not described in more detail here. This class could be assigned a function, such as refueling, etc.

```
"""Module car"""


# ----------------------------------------
# Auxiliary functions
# ----------------------------------------


def value_return(label, value, unit=none):
    """Formatted return of values with and without unit"""
    if unit is none:
        print('{b}: {w}'.format(b=label, w=value))
    elif unit == 'Euro' or unit == 'EUR':
        print('{b}: {w:.2f} {e}'.format(b=label, w=value,
                    e=unit))
    else:
        print('{b}: {w} {e}'.format(b=label, w=value, e=unit))


def fuelcosts_per_km(price_in_euro_per_l, consumption_in_l_per_100km):
    """Calculation of the fuel costs per kilometers

    :param price_in_euro_per_l: price per liter [EUR/l]
    :param consumption_in_l_per_100km: consumption of 100 km [l/100km]
    :return: fuel costs per kilometers [EUR/km]
    """
    return price_in_euro_per_l * consumption_in_l_per_100km / 100


# ----------------------------------------
# Defining class car
# ----------------------------------------


class car:
    """Each instance of the class car has a range."""


    def __init__(self, range):

        self.range = range                # range in km per tank filling
        self.amount_fuelling = 0          # amount of fuelling
        self.consumption_costs = 0        # accumulated consumption costs
        self.total_costs = 0              # accumulated total costs
```

```python
    def fuelling(self, costs_per_km):

        # Mileage (driven km) calculate and show
        mileage = self.amount_fuelling * self.range
        value_return('mileage', mileage, 'km')

        # Show previous consumption costs
        value_return(' consumption costs before fuelling',
                        self.consumption_costs, 'Euro')

        # Increase and show counter of fuelling
        self.amount_fuelling += 1
        value_return('amount of fuelling', self. amount_fuelling)

        # Calculate and return costs of fuelling
        costs_fuelling = self.range * costs_per_km
        Show_value(' costs of fuelling ', costs_fuelling, 'Euro')

        # Calculate new and show accumulated consumption costs and total
        # costs
        self.consumption_costs += costs_fuelling
        self.total_costs += costs_fuelling
        show_value('consumption costs after fuelling',
                        self.consumption_costs, 'Euro')
        show_value('total costs after fuelling', self.total_costs,
                        'Euro')

class burners(car):
    """ Each instance of the class burners can have a soot filter with a
        defined lifetime. ""

    def __init__(self, range, filter_lietime_km=None):
        super(burner, self).__init__(range)
        self.drive_technology = 'combustion engine'

        if filter_lifetime_km:
            self.filter_lifetime_km = filter_lifetime_km
            # Lifetime of a soot filter in km
            self.filter_km = 0
            # Driven km with soot filter

    def fuelling(self, costs_pro_km):
        super(burner, self).fuelling(costs_per_km)
```

```python
    if self.filter_km is not None:

        # Show mileage of the soot filter
        show_value(' mileage of the soot filter',
                    self.filter_km, 'km')

        # Check lifetime of the soot filter
        if self.filter_lifetime_km + self.range <=
                        self.filter_km:
            print('> Soot filter broken. Change!!! <')
        elif self.filter_lifetime_km <= self.filter_km:
            print('> Change soot filter! <')
        elif self.filter_lifetime_km <= self.filter_km +
                        self.range:
            print('>Soot filter has to be changed soon.<')

        # Calculate future mileage (next fuelling) of the soot filter:
        self.filter_km += self.range

def filter_change(self, costs_change,
                    filter_lifetime_km=None):
    """Change soot filter

    Always after fuelling the soot filter gets changed (condition for
    calculating the mileage of the soot filter. Alternative the
    mileage of the soot filter could also be calculated by the amount
    of fuellings, like the mileage of the car.
    """

    if filter_lifetime_km is not None:

        # Set new lifetime of the soot filters
        self.filter_lifetime_km = filter_lifetime_km

    if self.filter_lifetime_km is not None:

        # Set mileage (next fuelling) of the soot filter
        self.filter_km = self.range

        # Show previous consumption costs
        show_value('Consumption Costs bevor changing soot filter',
                    self.consumption_costs, 'Euro')
```

```
         # New calculation and show accumulated total costs
         self.total_costs += change_costs
         show_value('Total costs after changing soot filter',
                          self.total_costs, 'Euro')

class evehicle(car):

    def __init__(self, range):

         super(evehicle, self).__init__(range)
         self.drive_technology = 'Electric motor'
```

What result outputs does the program produce? In order to generate outputs, a file 'auto_run.py' is generated, which accesses the data files by means of the code shown above. This file is called 'auto.py' and must be contained in the same directory as the file 'auto_run.py'. This 'auto_run.py' file is accessed via another Python program, such as Spyder (cf. Sect. 4.3). By means of the command *from auto import* * all classes and methods are imported. After importing class properties, an object, such as 'convertible' can be created. For each object, the cost of fuel is defined (*convertible.fuelling(0.09)*). After running this program, the results are given out in Spyder in a separate console. In this example, the output values were added behind the hashtags as commentaries.

```
#
# Examples of calls on the console
#

from car import *
#
convertible = car(500)
#
convertible.fuelling(0.09)
# Mileage: 0 km
# Consumption costs before fuelling: 0.00 Euro
# Amount of fuelling: 1
# Fuel costs: 45.00 Euro
# Consumption costs after fuelling: 45.00 Euro
# Total costs after fuelling: 45.00 Euro
#
convertible.fuelling(0.1)
# Mileage: 500 km
# Consumption costs after fuelling: 45.00 Euro
# Amount of fuelling: 2
# Fuel costs: 50.00 Euro
```

```
# Consumption costs after fuelling: 95.00 Euro
# Total costs afer fuelling: 95.00 Euro
#
bus = burner(600, 2400)
#
bus.fuelling(0.25)
# Mileage: 0 km
# Consumption costs before fuelling: 0.00 Euro
# Amout of mileage: 1
# Fuel costs: 150.00 Euro
# Consumption costs after fuelling: 150.00 Euro
# Total costs after fuelling: 150.00 Euro
# Mileage of the soot filter: 0 km
#
bus.fuelling(0.25)
# Mileage: 600 km
# Consumption costs before fuelling: 150.00 Euro
# Amout of mileage: 2
# Fuel costs: 150.00 Euro
# Consumption costs after fuelling: 300.00 Euro
# Total costs after fuelling: 300.00 Euro
# Mileage of the soot filter: 600 km
#
bus.fuelling(0.25)
# Mileage: 1200 km
# Consumption costs before fuelling: 300.00 Euro
# Amout of mileage: 3
# Fuel costs: 150.00 Euro
# Consumption costs after fuelling: 450.00 Euro
# Total costs after fuelling: 450.00 Euro
# Mileage of the soot filter: 1200 km
#
bus.fuelling (0.25)
# Mileage: 1800 km
# Consumption costs before fuelling: 450.00 Euro
# Amount of fuelling: 4
# Fuel costs: 150.00 Euro
# Consumption costs after fuelling: 600.00 Euro
# Total costs after fuelling: 600.00 Euro
# Mileage of the soot filter: 1800 km
# > The soot filter has to be changed soon. <
#
bus.fuelling(0.25)
# Mileage: 2400 km
# Consumption costs before fuelling: 600.00 Euro
```

```
# Amount of fuelling: 5
# Fuel costs: 150.00 Euro
# Consumption costs after fuelling: 750.00 Euro
# Total costs after fuelling: 750.00 Euro
# Mileage of the soot filter: 2400 km
# >Change soot filter! <
#
bus.filter_change(500)
# Consumption costs before changing soot filter: 750.00 Euro
# Total costs after changing soot filter: 1250.00 Euro
#
bus.fuelling(0.24)
# Mileage: 3000 km
# Consumption costs before fuelling: 750.00 Euro
# Amount of fuelling: 6
# Fuel costs: 144.00 Euro
# Consumption costs after fuelling: 894.00 Euro
# Total costs after fuelling: 1394.00 Euro
# Mileage of the soot filter: 600 km
```

This is only one simple example. In this way very complex programs can be created to run even more complexs calculations in order to answer questions on energy supply systems.

2.2 Structure of the Generic Model in OEMOF

As was discussed earlier, the OEMOF system of programs is based on an object-relational structure. It uses classes and objects. Furthermore, the system of programs has a modular structure, composed of libraries, local modules, and packages. Currently OEMOF contains nine user-programmed libraries. These libraries can take on specific tasks, such as the generation of performance graphs (load profiles) based on wind energy data. Some libraries support the running of specific modeling steps. Once an optimization program has been run and produced a solution, the next step is how to provide the output. The output library oemof. outputlib is available to support this step. In the OEMOF project, libraries are organized in functional units that are represented by different layers. These layers help to define the library contents (cf. Fig. 2.4).

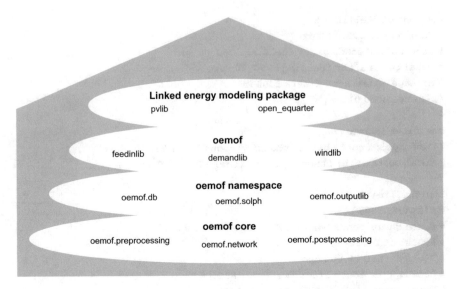

Fig. 2.4 Organization of OEMOF libraries in different layers (based on Hilpert 2017)

The four layers can be described in the following way:

– Core layer
   The core layer contains the core libraries with core classes relevant to all models
   and defined by the main interface. The core layer also contains the generic graph
   structures. An energy system as interpreted by OEMOF is described in the core
   layer libraries and the basic application programming interfaces (APIs) are
   defined in the core layer libraries. Furthermore, the input and output formats are
   defined in this libraries. This layer provides a basic structure for the graphic
   representation of an energy supply system.
– Namespace layer
   This is the layer for executing the optimization. This layer contains packages
   inherited from the basic modules and derived on their basis. They are connected to
   the core layer via APIs. An example of this might be libraries for optimization
   models for costs or energy flows. This layer is also called the modeling layer, as
   this is the storage place for models derived from previous classes and inheriting
   from these predefined classes. In the OEMOF version v0.2.0 the only library in
   this layer is the *solph* library. But in the future, other libraries might be added here.
– OEMOF layer (OEMOF cosmos)
   This layer contains libraries associated with OEMOF. This layer contains
   methods that can be used in OEMOF. But these methods can also be used in
   other contexts. These libraries are useful additions for modeling in OEMOF.
   This layer contains libraries containing programs and tools for energy optimization functions that are not system models and are not based on a graph
   structure. This means that they do not refer to the core classes and can be used

independently. However, they follow the same structure as the core libraries in order to facilitate using these programs.

– OEMOF-cosmos layer

This outer layer stores external packages of completed energy analysis projects that use the OEMOF open source programs. These can make use of libraries of the lower layers or create their own libraries. This enables synergies between previously created objects and allows future modelers to make use of earlier model samples. An example of such an external package is the PV library. This is of such a high quality that OEMOF has not created its own PV library.

An essential element of the libraries is descriptions of the model according to the network principle in the libraries *oemof.network* and *oemof.solph*. Based on the object-relational approach several core classes have been defined according to a hierarchical structure. Figure 2.5 describes this structure in the Unified Modeling Language (UML).

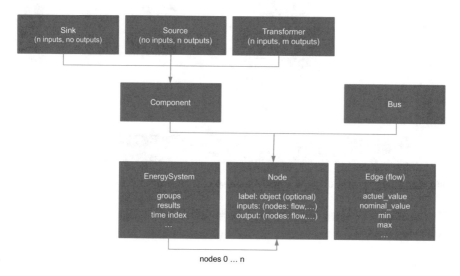

Fig. 2.5 Hierarchical Structure of Core Classes in OEMOF (UML) (based on Hilpert 2017)

At the top there are the core classes *EnergySystem* and *Flow* (these are called edges in graph theory and *nodes* as parent classes). Above these are the child classes that inherit from the parent classes (cf. Fig. 2.5). The class *EnergySystem* acts as a container. This means that user-specific information can be integrated into this class. An example for this might be timeframe data or data as to the groups included in the data. Groups are normally created according to specific criteria, such as the group of all transformers of the type gas turbines.

Modelers can also include user-defined features such as their specific system of inheritance. These features can be stored in the module *custom.py*. If these classes have been created correctly, have been tested and are free from programming mistakes, they will be stored in the module *components.py*.

From the external libraries together with the OEMOF specific libraries individual models can be created that can be used for specific applications. How a library can be linked to a specific application, can be seen in the following diagram (cf. Fig. 2.6).

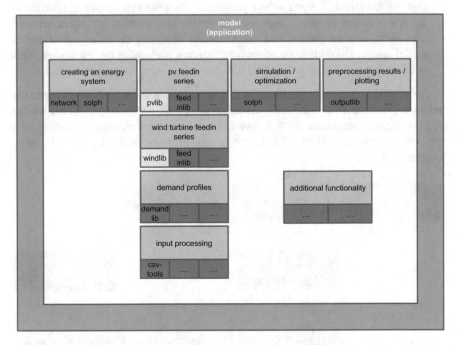

Fig. 2.6 Using libraries to create a model for a specific application (basied on Hilpert 2017)

The libraries shown in the figure above (cf. Fig. 2.6) will be discussed in the following Sect. 2.3.

2.3 Libraries in OEMOF

The OEMOF system of programs provides five key libraries:

– oemof-network
– oemof-solph
– oemof-outputlib
– feedinlib
– demandlib

These will be introduced in more detail in the following section. Users can access the source code of OEMOF via the programming hub *GitHub*. *GitHub* is a platform for developers that was created specifically for large open source projects. Access to OEMOF is via the Internet platform shown in Fig. 2.7. The homepage

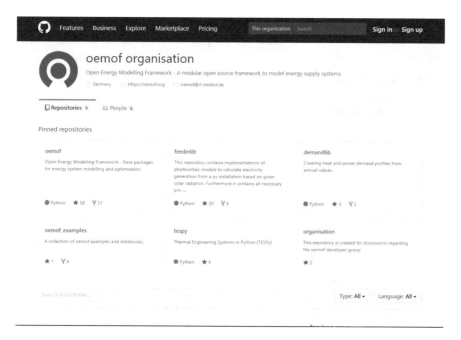

Fig. 2.7 OEMOF entry page at GitHub (GitHub 2017r)

gives the user an introduction to the structure of OEMOF. The homepage also links to the different library pages. The OEMOF specific libraries can be found when clicking on the OEMOF icon.

In a specific application, which we will here call an OEMOF application, the modeling of an energy supply system will normally require the use of one or more of the libraries listed here and may also require the use of additional, external libraries, in order to answer a specific question, such as the economic viability of extending the power-to-gas plant network.

2.3.1 The Oemof-Network Library

The oemof-network library is the basis for modeling an energy supply system in the oemof-solph library. When modeling an energy supply system, the oemof-network library is not directly accessed, but runs in the background. Via the oemof-solph library, the basic structure of the energy supply system model created by the modeler is then transformed into a graph structure based on graph theory.

The oemof-network library is automatically available to the modeler after completing the OEMOF installation (oemof 2014a).

This library contains the classes for describing key components of the energy supply system as well as the inputs and outputs required for the models (GitHub 2017a).

- class *Inputs(MM)*
- class *Outputs(MM)*
- class *_Edges(MM)*
- class *Node*
- class *Bus(Node)*
- class *Component(Node)*
- class *Sink(Component)*
- class *Source(Component)*
- class *Transformer(Component)*
- class *Entity*

Classes with brackets behind their name are child classes. The parent classes are given in the brackets after the child class. The parent classes contain the key methods, functions, parameters, attributes as well as other important data relevant to modeling. The abbreviation MM in brackets stands for MutableMapping. This is an abstract basic class (ABC) for container classes (Python 2018f). These abstract classes can be used to test whether classes provide a relevant interface and whether these are hashable (can be separated into smaller components) or if they are mapping classes. The ABC class *MutableMapping* provides the methods *__getitem__*, *__setitem__*, *__delitem__*, *__iter__* and *__len__*. These basic classes are imported via the command:

```
from collections import MutableMapping as MM
```

The class *Edges(MM)* represents the edges of the model in accordance with graph theory. The class *Entity* has not yet been discussed here. An entity is an abstract concept. Any type of element can be conceived as an entity in an energy supply system, an example might be a gas turbine or any other element. What is important is that each entity is identifiable and is connected to at least one other entity via an input or an output. The class *Entity* serves to pool the properties of different entities in order to make them available to future classes. This class is more precisely defined by the objects that the modelers create in their applications.

Within the different classes certain methods as well as parameters and attributes may be defined. These will be introduced in more detail in the following chapters. This library is stored in the files *network.py* (GitHub 2017a).

2.3.2 The Oemof-Solph Library

The oemof-solph library was developed in order to enable to create and solve mixed-integer linear optimization functions (oemof 2014a). These were created in the Pyomo programming language. Pyomo is a program package based on Python, which is also contained in the OEMOF installation package (oemof 2014a). Pyomo is an open source modeling tool which provides classes and functions designed to create optimization functions via an API. The Pyomo package contains a set of different optimization options.

Using the four basic classes *sink, source, transformer* and *bus* as well as other user-specific components, the modelers can model their energy supply systems. The components of an energy supply system are assigned to these classes in accordance with their previously defined properties.

The four basic classes represent the nodes of the model. The energy flows that are generated by energy inputs and outputs represent the edges of the model. In order to model energy flows, a fifth basic class needs to be created, the class *flow*. A specific OEMOF convention allows to represent this model as a grid plan, which was introduced in Chap. 1.2 and shown in Fig. 1.4. In this example, the classes *source* and *sink* are represented as chevrons, the class *transformer* as an oblong and the class *bus* as a circle, the user-specific class *store* is represented as a star. The connecting lines represent the flow, which is assigned a value, e.g., energy costs.

In the following section, we will have a look at a rather complex model, the energy supply system of the city of Rostock (this will be referred to as Rostock in the following pages). The model describes the energy supply system within and between two regions (region 1 and region 2) in the metropolitan area of Rostock, a city in the German state of Mecklenburg-Pomerania. The energy supply of each region is composed of several different components (cf. Table 2.3). The two regions are furthermore linked by an electricity cable. This cable allows electricity to flow from region 1 to region 2 and vice versa. The transport of energy and energy inputs and outputs are represented by buses. A bus is a financial node that must be balanced at any point in time. This means that a bus must never have a negative balance. The components and flows are listed in Table 2.3.

The energy grid following the OEMOF conventions can be seen in Fig. 2.8.

Table 2.3 Nodes and graphs representing an energy supply system, in this example Rostock, a city in the German state of Mecklenburg-Pomerania

Region 1	
Component	*Reference*
wind turbine	wt1
biogas-cogeneration plant	bg1
gas turbine	gt1
electrical demand	de1
heat demand	dh1
cable to region 2	cb1
Customer-specific component	*Reference*
storage	st1
Bus	*Reference*
gas_bus	r1_gas
electrical_bus	r1_el
local_heat_bus	r1_th
biogas_bus	r1_bio
Region 2	
Component	*Reference*
coal plant	cp2
chp plant (gas combined heat and power plant)	chp2
p2g-facility	ptg2
Electrical demand	de2
Cable to region 1	cb2
heat demand	dh2
Bus	*Reference*
electrical_bus	r2_el
local_heat_bus	r2_th
coal_bus	r2_coal
gas_bus	r2_gas

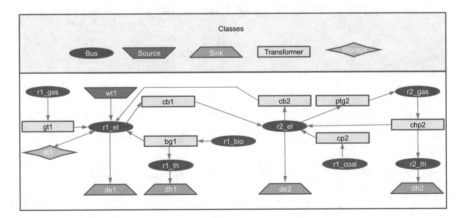

Fig. 2.8 Energy grid for Rostock

In order to model this in oemof-solph, this must be converted into a bipartite graph according to graph theory (cf. Fig. 2.9).

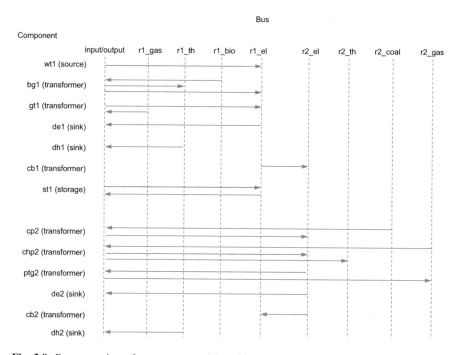

Fig. 2.9 Representation of an energy model as bipartite graph in OEMOF (based on oemof 2014c)

Following the graph model the model is then converted into a program by means of the source code. This structure is the entry information for the oemof-solph library. In order to create such a model, we must first define our components, such as the CHP and the links between these classes must be defined by the classes *flow* and *bus*.

A further requirement for creating the model is a container module, into which the specific energy model can be integrated. For this, modelers can make use of the module oemof.energy_system (file *energy_system.py*). This module defines the class *EnergySystem*. This class also contains several methods which can be executed within the instance of the class *EnergySystem* within the modelers' specific application (GitHub 2017b). One such instance of the class *EnergySystem* is Rostock, our example.

Before we can create our energy supply system model, we must first import the classes for the energy supply components from oemof.network and we must supply methods for the data processing of an energy system by means of the oemof.energy_system module. In order to do this, we must follow these programming steps:

```
# Certain classes from oemof.network are imported that are used in the own
# application, for example Sink, Transformer.

from oemof.network import Sink, Transformer, Bus, Flow

# The class EnergySystem can be installed together with further classes
# using the package oemof.solph:

from oemof.solph import (EnergySystem, Sink, Transformer, Bus, Flow, Model,
                         components)

# Once the class EnergySystem has been imported into the application, the
# program must be told how to call up the class
# in the following code. From then on, the EnergySystem can be
# accessed via the variable 'es'. The parameter 'datetimeindex' must be
# defined accordingly in Pandas in the class 'DatetimeIndex'
# (cf. Ch. 2.5).

es = EnergySystem(timeindex=datetimeindex)
```

In order to ensure that the structure of the energy supply system is entered into the oemof-solph library, the structure of the energy supply system must be converted to programming code.

```
# As an example for the infrastructure of an energy supply system,
# a Bus and a Sink are created.
# The term 'label' provides the bus with a name for representation in a
# graph here the name is 'b_gas'. The string before =
# is the term which will be used to access this component in the
# source code in future.

# Create bus 1

r1_gas = Bus(label='b_gas')

# The following section describes the implementation of a sink. This
# requires information about the input, in this case from Bus 1. This input
# is accessed via the bus 'r1_gas'.
# This input is further defined by the cost for the energy flow (Flow)
# given as a variable cost at 0,1 €/kWh. These costs of 0,1 €/kWh will be
# factored into the optimization function.
# For the overall modeling, it is important to note that parameters must
# always be calculated in the same unit, either € or $. The units defined
# for an input must be documented in the source code. Using different units
# for calculations tends to be one of the main reasons for incongruous
# modeling results. The biggest danger might be that these discrepancy
```

```
# might not even be noticed after the calculations have taken place.

# Create sink 1.

de1 = Sink(label='strom', inputs={r1_gas: Flow(variable_costs=0,1)})
```

In the code we can see that costs will be incurred. The class *Flow* represents the energy flows of an energy supply system. The library oemof.solph offers several predefined parameters for this class contained in the module 'network.py'. Some of these parameters deal with costs. But there are also parameters covering efficiency or defining limits, such as maximum yield.

But before oemof-solph can be implemented, at least one linear solver must be installed. So far, the oemof-solph library has been tested with the two solvers CMC and Gurobi. CBS is an open source solver, while Gurobi is a commercial solver, where the options for further development and reuse and adaptation by users are limited. The same is true for the solver Cplex. Another solver that can work with OEMOF models is GLPK.

Modelers can use oemof-solph to model an energy supply system using *oemof-network* as a reference.

2.3.3 The Oemof-Outputlib Library

This library can be used to present the results of an optimization function. The main purpose of this library is the summarization and presentation of results. This library is also part of the OEMOF program installation. This library is based on the Python-based library Pandas. Pandas is a tool for data analysis and presentation (Durmus 2017). Pandas also offers specific data structures and analysis tools. The results from the optimization function created in OEMOF are collected in a Pandas MultiIndex DataFrame. For this purpose data is transformed into a dictionary. This dictionary contains Pandas DataFrames and Pandas series for all nodes and flows. A DataFrame contains scalar data (e.g., investments) and sequences describing nodes (with keys like (node, None)) and flows between nodes (with keys like (node_1, node_2)). You can directly extract the dataframes into the dictionary by using these keys, where 'node' is the name of the object you want to address. If you want to address objects by their label you can convert the results dictionary by changing the keys into strings given by the labels (GitHub 2017a):

```
views.convert_keys_to_strings(results)
print(results[(wind, bus_electricity)]['sequences']
```

Another option is to access data belonging to a grouping by the name of the grouping (cf. Sect. 2.4.2). Given the label of an object, e.g., 'wind', you can access the grouping by its label and use this to extract data from the results dictionary.

```
node_wind = energysystem.groups['wind']
print(results[(node_wind, bus_electricity)])
```

Further information can be found at: (pandas o.J.-a).

Oemof-outputlib also provides further plot methods for creating graphs and diagrams. See also (oemof 2014a).

As an example we present an excerpt from the Rostock source code, which represents the collection of results in a Pandas DataFrame. In a next step, the data of one specific component are selected. In this example, the data for the heat component is selected. The data is collected by means of the processing module. You can see this data filter at work in the following excerpt.

```
results = outputlib.processing.results(om)
data_heat = outputlib.views.node(results, 'heat')
```

It can also be important to be able to select data for a specific node. This is what the module 'view' is designed for. The second line of code presents a query eliciting all relevant data for this node, requesting either the variable for the node or its label.

Another function allows for the collection and printing of metadata, i.e., information about the function or the optimization problem or the solver.

```
meta_results = outputlib.processing.meta_results(om)
pp.pprint(meta_results)
```

The result of this query can then be visually displayed by means of the Spyder system of programs, if the created modeling code is entered into this program.

As an example of a graphic output, see Fig. 2.10.

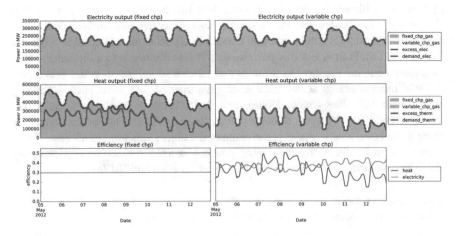

Fig. 2.10 Graphic representation of data on wind energy (oemof o.J.-b)

2.3.4 The Feedinlib Library

This library is not part of the OEMOF installation. It is installed via PyPI. The abbreviation stands for Python Package Index (Python o.J.-a). This is a storage place for Python packages. PyPI can be accessed via the pip program. Pip is the package manager for installing Python packages (Python 2018b).

The feedinlib library can be used to compute output data for wind power and solar power stations (oemof 2014a). Feedinlib provides technical parameters for wind and solar power stations and additionally provides a weather dataset for wind and sun hours as a basis for the calculations. For both types of power stations temperature data is required to run the program. For the calculation of the wind power functions, the wind speed, the atmospheric pressure and the roughness length is also required. For the solar power load profile, the required data includes information on direct sunlight and sunlight diffusion. The modelers can customize some of the data to suit their data. The output of these computations is a load profile for the wind or solar power stations.

When you install the feedinlib library, you receive a CSV file with the relevant data for wind and solar power. The CSV file for wind power stations contains a list of more than 90 types of wind power stations. The list also includes the power coefficient c_P. For wind power calculations, the performance is described by means of the power coefficient. The reason for this calculation basis is the fact that wind power is related to aerospace technology, where the drag and lift coefficients are used.

The power coefficient c_P can be derived from the efficiency value η. The power coefficient c_P is based on Albert Betz' theory from 1920 (Betz 1994). The efficiency value η is calculated based on the ratio of energy consumed by a machine and energy or performance produced by that machine. The mathematical formula for this is (cf. Eq. 2.1):

$$\eta = \frac{P_{out}}{P_{in}} \tag{2.1}$$

Based on

$P_{out} = P_{Rotor}$ Wind energy or usable energy of the energy produced by the rotor blades corresponds

$P_{in} = P_{Wind}$ To the maximum energy that a wind power station can yield

The power coefficient c_P for wind turbines is calculated according to Eq. 2.2 (Ratka 2015; Quaschning 2015):

$$\eta = c_P = \frac{P_{Rotor}}{P_{Wind}} \tag{2.2}$$

If the total yield of a wind power station is to be calculated, the rotor performance is replaced by the total electricity yield of a wind power station that can be fed back to the grid. The power that can be fed back to the grid is dependent on the wind speed. That is why the following equation is used (cf. Eq. 2.3) (Quaschning 2015):

$$P_{\text{Rotor}} = P = P(v_{\text{Wind}}) \tag{2.3}$$

Based on

v_{Wind} Wind speed
P Electricity yield that is fed back to the grid

The yield based on wind speed is calculated according to Eq. 2.4 (Quaschning 2015):

$$P_{\text{Wind}}(v_{\text{Wind}}) = F_{\text{Wind}} \cdot v_{\text{Wind}} = \frac{\rho_{\text{Air}}}{2} \cdot A \cdot v_{\text{Wind}}^3 \tag{2.4}$$

Based on

F_{Wind} Wind power affecting the swept rotor area perpendicular to the wind direction. This force depends on air density and the rotor area it affects.
ρ_{Air} Air density
A Area affected by the wind power. Swept rotor area perpendicular to the wind direction. The CSV file for solar modules contains data for more than 500 different modules.

The parameter set for solar modules as well as the cp-values for wind power stations can be downloaded from the OEMOF site at https://github.com/oemof/feedinlib. If you install the OEMOF feedinlib, you will find the files in the following directories:

- C:\WinPython-64bit-3.6.3.0Qt5\python-3.6.3.amd64\Lib\pydoc_data
- C:\WinPython-64bit-3.6.3.0Qt5\python-3.6.3.amd64\Lib\site-packages\wind-powerlib\data.

Of course, it is also possible to create your own list of files.

The following programming code allows you to start calculating the load profiles of a wind power station or of a solar array.

```
# Create object 'my_weather' of the class 'weather.FeedinWeather'.

my_weather = weather.FeedinWeather()

# Create function of reading data from the CSV file
# call up 'weather.csv' and provide the filename
# as parameter.
```

```
my_weather.read_feedinlib_csv(filename='weather.csv')

# Data of the wind power plant is read by means of the wind CSV file.
# in this example the object 'enerconE126'.

E126_power_plant = plants.WindPowerPlant(**enerconE126)

# This code creates the object 'E126_power_plant' and assigns it to the
# parent class 'plants.WindPowerPlant'.
# The string in brackets (**enerconE126) has the meaning: By means for the
# two asterisks ** more parameters can be defined by means of a dictionary.
# The dictionary is then dissolved into the string:
# 'key1=value1, key2=value2,...'

# Now the weather data is imported. The maximum power produced by the
# wind power stations is limited to 15 MW
E126_feedin = E126_power_plant.feedin(weather=my_weather,
                        installed_capacity=15000000)

# A solar power plant is created in the same way. In this example, the
# number of solar modules is limited to 5000 modules.

yingli_module = plants.Photovoltaic(**yingli205, YL205P-23b)
pv_feedin = yingli_module.feedin(weather=my_weather, number=50000)
# 50000 modules
```

2.3.5 The Library Demandlib

This is another library that is not a part of the automatic OEMOF installation package. It must be added via PyPI (oemof 2014a). The demandlib library creates electricity and heat load profiles from existing demand load profiles. These load profiles can be created for different sectors, such as private households or industrial production (oemof 2016). Load profiles can also be set in accordance with the modelers' requirements.

2.4 Key Packages and Modules in OEMOF

The OEMOF system of programs already provides some key packages to modelers wanting to create an optimization model for an energy supply system. These packages are modular in structure. The modules are stored in separate files. All these modules have the file ending *.py*. These modules define and configure the classes that can then be accessed by means of instantiating them in your source code.

The module energy_system has already been introduced. This can be found in the top layer of the OEMOF hierarchy. This is also where we find the modules *groupings*, which groups entities together in relevant groupings.

In the next layer below the top layer we find three key module packages. These key modules include (oemof 2014e):

- oemof.outputlib package
- oemof.solph package
- oemof.tools package

The following section will introduce the module *energy_system* followed by the module *groupings*, before presenting the three key packages.

2.4.1 Module Energy_System

The first step in using OEMOF to model an energy system is to create and define your energy system. This is the prerequisite to using the OEMOF solver libraries. To do this OEMOF makes the module oemof.energy_system (oemof.energy_system) available. This module defines the class *EnergySystem* (*oemof.energy_system. EnergieSystem*):

```
class EnergySystem(**kwargs)
```

This class inherits from the parent class *object*.

The parameter *kwargs* represents *keyword arguments*. This parameter is a placeholder parameter that can be replaced by any keyword arguments the user wants to define.

Several parameters are available for modeling in the class *EnergySystem* (cf. Table 2.4). The parameters are written into the attributes for the class *EnergySystem*.

The attributes for this class are listed in Table 2.5.

Table 2.4 List of parameters for the module *oemof.energy_system* (GitHub 2017b)

Parameter	Additional information	Description
entities	– List of entity – :class:'Entity <oemof.core. network.Entity>' – Optional	A list containing the already existing entities that should be part of the energy system. Stored in the entities attribute. Defaults to [] if not supplied. If this EnergySystem (:class:'EnergySystem') is set as the registry attribute (:attr:'registry <oemof.core.network.Entity.registry>'), which is done automatically on EnergySystem construction, newly created entities (:class:'Entities <oemof.core.network.Entity>') are automatically added to this list on construction.
timeindex	pandas.index, optional	Define the time range and increment for the energy system. This is an optional parameter but might be important for other functions/methods that use the EnergySystem class as an input parameter.
groupings	list	The elements of this list are used to construct *groupings* or they are used directly if they are instances of grouping. These groupings are then used to aggregate the entities added to this energy system into groups. By default, there'll always be one group for each UID (unified identifier) containing exactly the entity with the given uid.

Table 2.5   Attributes of the class *EnergySystem* (GitHub 2017b)

Attribute	Additional information	Description
entities	list of :class:'Entity <oemof.core. network.Entity> '	A list containing the : class:'Entities <emof.core.network.Entity>' that comprise the energy system. If this :class:'EnergySystem') is set as the attribute (:attr:'registry <oemof.core.network.Entity. registry>'), which is done automatically on : class: construction, newly created :class: 'Entities <oemof.core.network.Entity>') are automatically added to this list on construction.
groups	Data type: dict	
results	Data type: dictionary	A dictionary holding the results produced by the energy system. Is 'None' while no results are produced. Currently only set after a call to : meth:'optimize' after which it holds the return value of :meth: 'om.results() <oemof.solph. optimization_model.OptimizationModel. results>'. See the documentation of that method for a detailed description of the structure of the results dictionary.
timeindex	pandas.index, optional	Define the time range and increment for the energy system. This is an optional attribute but might be important for other functions/ methods that use the EnergySystem class as an input parameter.

2.4.2   Module Groupings

Another important module is *oemof.groupings*. By means of this module the entities of an energy system can be grouped together. For this to happen, first the class *Grouping* must be defined. This class is called whenever a new entity is added to an energy supply system. In OEMOF, every node that is added to a model is assigned a group. This makes it possible for a specific group to carry out joint activities, such as setting constraints, etc. These entities are placed within the module groupings and assigned a unified identifier (UID). If an entity is placed in the groupings module and there is already an entity with the same UID, an error message is produced. The following string of code is created within the source code.

```
oemof.groupings.DEFAULT = <oemof.groupings.Grouping object>
```

The string *oemof.groupings.DEFAULT* defines an object of the *Grouping* class. This is a placeholder object, a default object, which ensures that each note represents a separate group.

Entities of this type are assigned the attribute *groups*. This is represented simply as a *g* in the code. The class is called via:

```
# The string 'g' represents the class Groupings, 'e' represents the class
# 'Entity', 'groups' is the attribute of the class 'e'. The attribute
# 'groups' is a dictionary, which contains the key for the groups.
g(e, groups)
```

This module contains three child classes, also called subclasses. The parent class is always given in brackets.

1. class *oemof.grouping*—class *Grouping*
2. class *oemof.groupings.Nodes*—class *Nodes(Groupings)*
3. class *oemof.groupings.Flows*—class *Flow(Nodes)*
4. class *oemof.groupings.FlowsWithNodes*—class *FlowsWithNodes(Nodes)*

The class *Grouping* is the parent class of the class *Nodes*. This class is the parent class for the classes *Flow* and *FlowsWithNodes*. The class *Grouping* is called whenever an entity is added to the *EnergySystem*. There are specific parameters for this class as listed in Table 2.6.

Table 2.6 List of parameters for the class *Grouping* (GitHub 2017c)

Parameter	Additional information	Description
key(e)	callable or hashable	This is the key ID under which this group is stored. This parameter is assigned to each entity *e* in the energy system. If this parameter is not callable, then a key for each entity of the energy system is defined. If the parameter is callable, the key is removed from each entity in the energy system. If an entity is not stored in a grouping, the value *None* must be the output.
constant_key	(hashable, optional)	A constant key is created.
value	(callable, optional)	This parameter overwrites any existing default value.
filter	(callable, optional)	If this parameter is assigned, any output *value()* is filtered accordingly.
merge (new, old)	(callable, optional)	This parameter indicates that an existing grouping is to be merged with the new grouping. This method is called if a value has already been assigned to *group[key(e)]*. If this happens, the function *merge(value(e), group[key(e)])* a is called. This will store the new grouping under parameter *key(e)*.

Other classes in this module group certain entities together. Let us now list these classes:

First the class *oemof.groupings.Nodes*.

This class is used to group nodes. This class is also designed to create sets of entities. The class is based on *oemof.groupings.Grouping*. This is called by the following string of source code:

```
class oemof.groupings.Nodes(key=None, constant_key=None, filter=None,
**kwargs)
```

The strings in brackets are filters for the key parameter *constant_key*.

Second, the class *oemof.groupings.Flows*.

This class groups all flows into sets that are connected to a node. For this reason, the parent class for these is the class *oemof.groupings.Nodes*. Of course, the class *oemof.groupings.Grouping* must receive the required information. Once the flows have been grouped, the functions *key* and *value* are applied to the set of flows.

The class *oemof.groupings.Flows* is defined in the OEMOF source code in the following way:

```
class oemof.groupings.Flows(key=None, constant_key=None, filter=None,
**kwargs)
```

The result of the grouping is a set of flows.

Thirdly, more information on the class *oemof.groupings.FlowsWithNodes*.

This class groups the nodes with their flows. This means that a set of tuples is created combining the values *source, target* and *flow*. Mathematically, a tuple is a list of and infinite amount of objects. These objects may or may not be separate entities. But the order of objects in a tuple is very important. From a mathematical point of view, tuples are sets with a specific order.

For this reason, the functions *key* and *value* are executed in a tuple.

In the programming code, this produces the following string.

```
class oemof.groupings.FlowsWithNodes(key=None, constant_key=None,
filter=None, **kwargs)
```

This means that a tuple set is produced.

2.4.3 The *oemof.outputlib* Package

This OEMOF package includes two modules:

- *processing.py*
- *views.py*

These two modules provide classes, parameters, and functions for creating graphics. This allows the modelers to not only create a model of their energy supply system but also to create a graphic representation of the results.

2.4.4   The oemof.solph Package

This is a key package, as this is the package in which the optimization model is created. This package contains the modules (GitHub 2017d):

– *blocks.py*
– *components.py*
– *constraints.py*
– *custom.py*
– *facades.py*
– *groupings.py*
– *models.py*
– *network.py*
– *options.py*

Contents of the module blocks.py

Blocks represent blocks of data. Blocks combine sets, variables, constraints, and aspects of the objective function for specific groups. There is always one main class and one block class for each component. The main class initializes the component. The block class is used to provide the constraints for each component, i.e., their mathematical description. For the core modules, both classes are available in two separate modules, in *solph.network* and in *solph.blocks*, which is due to the system of inheritance. For other components, these two modules are listed one after the other.

There are data blocks for the following classes:

– class *Flow(SimpleBlock)*
– class *InvestmentFlow(SimpleBlock)*
– class *Bus(SimpleBlock)*
– class *Transformer(SimpleBlock)*
– class *NonConvexFlow(SimpleBlock)*

All these listed classes inherit from the class *SimpleBlock*. This class can be found in the Pyomo package. In order to fully define a block, there are four classes of blocks:

– *BlockData*
– *Block*
– *SimpleBlock*
– *IndexedBlock*

The class *BlockData* contains an exact description of the block as a system. This is the container which includes all components and all relevant data to implement the block and to connect to interfaces. *SimpleBlock* and *IndexedBlock* are the components, that connect to other blocks. The class *Block* contains all methods, such as the method '__new__()'. This method allows for the transformation of the class *Block* into the class *SimpleBlock* or *IndexedBlock* based on a constructor argument.

For the first class *Flow* the following variables are made available (cf. Table 2.7):

Table 2.7 Variables for the class *Flow* (GitHub 2017k)

Variable	Description
negative_gradient	Difference of a flow in consecutive timesteps if flow is reduced indexed by NEGATIVE_GRADIENT_FLOWS, TIMESTEPS.
positive_gradient	Difference of a flow in consecutive timesteps if flow is increased indexed by POSITIVE_GRADIENT_FLOWS, TIMESTEPS.

For this class, for all flows that have a time limit the following sets are created (cf. Table 2.8):

Table 2.8 Sets for the class *Flow* (GitHub 2017k)

Set	Description
SUMMED_MAX_FLOWS	A set of flows with the attribute :attr:'summed_max' being not None.
SUMMED_MIN_FLOWS	A set of flows with the attribute :attr:'summed_min' being not None.
NEGATIVE_GRADIENT_FLOWS	A set of flows with the attribute : attr:'negative_gradient' being not None.
POSITIVE_GRADIENT_FLOWS	A set of flows with the attribute : attr:'positive_gradient' being not None.
INTEGER_FLOWS	A set of flows where the attribute :attr:'integer' is True (forces flow to only take integer values).

The mathematical description for calculations will be given in Chap. 3.

By means of the parameter *groups,* the objects in this class can be grouped (cf. Table 2.9).

Table 2.9 Parameter of the class *Flow* (GitHub 2017k)

Parameter	Data type	Additional information
group	list	List containing tuples containing flow (f) objects and the associated source (s) and target (t) of flow, e.g., groups = [(s1, t1, f1), (s2, t2, f2), …].

The next class is *class InvestmentFlow(SimpleBlock)*. This block gives information on the required investment, which requires for the value of the attribute *investment* not to be None. There are a number of sets for this class (cf. Table 2.10):

Table 2.10 Sets for the class *InvestmentFlow* (GitHub 2017k)

Set	Description
FLOWS	A set of flows with the attribute :attr:'invest' of type :class:'. options.Investment'.
FIXED_FLOWS	A set of flows with the attribute :attr:'fixed' set to 'True'.
SUMMED_MAX_FLOWS	A subset of set FLOWS with flows with the attribute : attr:'summed_max' being not None.
SUMMED_MIN_FLOWS	A subset of set FLOWS with flows with the attribute :attr: 'summed_min' being not None.
MIN_FLOWS	A subset of FLOWS with flows having set a value of not None in the first timestep.

The following variable is also important (cf. Table 2.11).

Table 2.11 Variable of the class *InvestmentFlow* (GitHub 2017k)

Variable	Description
invest :attr: 'om. InvestmentFlow.invest[i, o]'	Value of the investment variable (i.e., equivalent to the nominal value of the flows after optimization (indexed by FLOWS)).

A grouping parameter is also available for this class (cf. Table 2.12):

Table 2.12 Parameter for the class *InvestmentFlow* (GitHub 2017k)

Parameter	Data type	Additional information
group	list	List containing tuples containing flow (f) objects that have an attribute investment and the associated source (s) and target (t) of flow, e.g., groups = [(s1, t1, f1), (s2, t2, f2), ...].

For the class *Bus(SimpleBlock)* another parameter is listed here (cf. Table 2.13):

Table 2.13 Parameter of the class *Bus* (GitHub 2017k)

Parameter	Data type	Additional information
group	list	List of oemof bus (b) object for which the bus balance is created, e.g., group = [b1, b2, b3, ...].

This class includes all data for buses with a balanced input and output.

The penultimate class is the class *Transformer(SimpleBlock)*. This class includes all data for linear connections between nodes of the type transformer together with the relevant set (cf. Table 2.14).

Table 2.14 Set for the class *Transformer* (GitHub 2017k)

Set	Description
TRANSFORMERS	A set with all :class:'~ oemof.solph.network.Transformer' objects.

This class again contains a parameter for groupings (cf. Table 2.14) (Table 2.15).

Table 2.15 Parameter for the class *Transformer* (GitHub 2017k)

Parameter	Data type	Additional information
group	list	List of oemof.solph.Transformers objects for which the linear relation of inputs and outputs is created, e.g., group = [trsf1, trsf2, trsf3, …]. Note that the relation is created for all existing relations of all inputs and all outputs of the transformer. The components inside the list need to hold an attribute 'conversion_factors' of type dict containing the conversion factors for all inputs to outputs.

The final class *NonConvexFlow(SimpleBlock)* also defines sets (cf. Table 2.16): For this class all flows must have the attribute *Nonconvex*.

Table 2.16 Sets for the class *NonConvexFlow* (GitHub 2017k)

Set	Description
MIN_FLOWS	A subset of set NONCONVEX_FLOWS with the attribute :attr:'min' beeing not None in the first timestep.
STARTUP_FLOWS	A subset of set NONCONVEX_FLOWS with the attribute :attr:'startup_costs' being not None.
SHUTDOWN_FLOWS	A subset of set NONCONVEX_FLOWS with the attribute :attr:'shutdown_costs' being not None.

For this class there is also a grouping parameter (cf. Table 2.17).

Table 2.17 Parameter of the class *NonConvexFlow* (GitHub 2017k)

Parameter	Data type	Additional information
group	list	List of oemof.solph.NonConvexFlow objects for which the constraints are build.

Key variables for this class are listed in Table 2.18.

Table 2.18 Variables for the class *NonConvexFlow* (GitHub 2017k)

Variable	Data type	Attribute	Description
status	binary	attr:'om. NonConvexFlow. status'	Variable indicating if flow is >= 0 indexed by FLOWS.
startup	binary	:attr:'om. NonConvexFlow. startup'	Variable indicating startup of flow (component) indexed by STARTUP_FLOWS.
shutdown	binary	:attr:'om. NonConvexFlow. shutdown'	Variable indicating shutdown of flow (component) indexed by SHUTDOWN_FLOWS.

Contents of the module components.py

In this module the user-specific components and classes together with their con-straints (associated individual constraints) are stored as well as information about user-specific groupings. For this class, there is also a block class for data blocks. This contains the required constraints for each class. This means that all information for the classes and their constraints is stored in the same place.

So far, the following classes have been defined within this module:

- *GenericStorage(network.Transformer)*
- GenericStorageBlock(SimpleBlock)
- *GenericInvestmentStorageBlock(SimpleBlock)*
- *GenericCHP(network.Transformer)*
- *GenericCHPBlock(SimpleBlock)*
- *ExtractionTurbineCHP(solph_Transformer)*
- *ExtractionTurbineCHPBlock(SimpleBlock)*

The first class describes storage modules in general. The class *Storage* inherits from the class *network.Tranformer*.

class GenericStorage(network.Transformer)

The class *GenericStorage* models key aspects of a storage system. This class has a number of parameters (cf. Table 2.19).

Table 2.19 Parameters of the class *GenericStorage* (GitHub 2017e)

Parameter	Data type	Additional information
nominal_capacity	numeric	Absolute nominal capacity of the storage.
nominal_output_capacity_ratio	numeric	Ratio between the nominal outflow of the storage and its capacity. For batteries this is also known as c-rate. Note: This ratio is used to create the Flow object for the outflow and set its nominal value of the storage in the constructor. If no investment object is defined it is also possible to set the nominal value of the flow directly in its constructor.
nominal_input_capacity_ratio	numeric	Ratio between the nominal inflow of the storage and its capacity. see: nominal_output_capacity_ratio.
initial_capacity	numeric	The capacity of the storage in the first (and last) time step of optimization.
capacity_loss	numeric (sequence or scalar)	The relative loss of the storage capacity from between two consecutive timesteps.
inflow_conversion_factor	numeric (sequence or scalar)	The relative conversion factor, i.e., efficiency associated with the inflow of the storage.
outflow_conversion_factor	numeric (sequence or scalar)	see: inflow_conversion_factor.
capacity_min	numeric (sequence or scalar)	The nominal minimum capacity of the storage as fraction of the nominal capacity (between 0 and 1, default: 0). To set different values in every time step use a sequence.
capacity_max	numeric (sequence or scalar)	see: capacity_min.
investment		:class:'oemof.solph.options.Investment' object Object indicating if a nominal_value of the flow is determined by the optimization problem. Note: This will refer all attributes to an investment variable instead of to the nominal_capacity. The nominal_capacity should not be set (or set to None) if an investment object is used.

The next class is *GenericStorageBlock*. Some mathematical constraints have been defined for this class. These are listed in Chap. 3.

A specification of the class *GenericStorageBlock* is represented by the class *GenericInvestmentStorageBlock*. This class inherits from *SimpleBlock*.

  class GenericInvestmentStorageBlock(SimpleBlock)

The following sets belong to this class (cf. Table 2.20).

Table 2.20   Sets for the class *GenericInvestmentStorageBlock* (GitHub 2017e)

Set	Description
INVESTSTORAGES	A set with all storages containing an Investment object.
INITIAL_CAPACITY	A subset of the set INVESTSTORAGES where elements of the set have an initial_capacity attribute.
MIN_INVESTSTORAGES	A subset of INVESTSTORAGES where elements of the set have a capacity_min attribute greater than zero for at least one timestep.

Sets are created in order to define constraints relevant for all components contained in this set.

Furthermore, the following variables can be used (cf. Table 2.21).

Table 2.21   Variables for the class *GenericInvestmentStorageBlock* (GitHub 2017e)

Variable	Attribute	Description
capacity	om.InvestmentStorage.capacity [n, t]	Level of the storage (indexed by STORAGES and TIMESTEPS).
invest	om.InvestmentStorage.invest [n, t]	Nominal capacity of the storage (indexed by STORAGES).

The parameter *n* represents a component and *t* represents the time.

The next class *GenericCHP* describes a generic Combined Heat and Power (CHP) component. This class inherits from the parent class *Network.Transformer*:

  class GenericCHP(Network.Transformer)

This class can be used to model back-pressure turbines, combined cycle extraction or back-pressure turbines. Power plants can be modeled mathematically by using a mixed-integer linear approach. For modeling these, a number of key parameters must be defined (cf. Table 2.22).

Table 2.22 Parameters of the class *GenericCHP* (GitHub 2017e)

Parameter	Data type	Additional information
fuel_input	dict	Dictionary with key-value-pair of 'oemof.Bus' and 'oemof.Flow' object for the fuel input.
electrical_output	dict	Dictionary with key-value-pair of 'oemof.Bus' and 'oemof.Flow' object for the electrical output. Related parameters like 'P_max_woDH' are passed as attributes of the 'oemof.Flow' object.
heat_output	dict	Dictionary with key-value-pair of 'oemof.Bus' and 'oemof.Flow' object for the heat output. Related parameters like 'Q_CW_min' are passed as attributes of the 'oemof.Flow' object.
beta	list of numerical values	Beta values in same dimension as all other parameters (length of optimization period).
back_pressure	boolean	Flag to use back-pressure characteristics. Works of set to 'True' and 'Q_CW_min' set to zero.

The linear connection between nodes of the class *GenericCHP* is modeled by means of the class *GenericCHPBlock*. This is a data block. The parent class for this is:

class GenericCHPBlock(SimpleBlock)

This class contains a parameter for grouping CHP objects (cf. Table 2.23).

Table 2.23 Parameters of the class *GenericCHPBlock* (GitHub 2017e)

Parameter	Data type	Additional information
group	list	List containing 'GenericCHP' objects. e.g., groups=[ghcp1, gchp2, …].

Severable variables are defined for the creation of the mathematical functions for this class, as shown in Table 2.24.

A list of constraints is also included, more information on this will be given in Chap. 3.

The class *ExtractionTurbineCHP* is available in order to model extraction turbines based on a linear mathematical approach. The core class this is based on is *solph_Transformer*:

class ExtractionTurbineCHP(solph_Transformer)

Table 2.24 Variables of the class *GenericCHPBlock* (GitHub 2017e)

Variable	Description	Nomenclature (Mollenhauer 2016)
self.H_F	var(self.GENERICCHPS, m. TIMESTEPS, within=NonNegativeReals)	fuel input $\dot{H}_F$: is usually calculated based on the lower heating value of the fuel and the corresponding mass flow.
self. H_L_FG_max	var(self.GENERICCHPS, m. TIMESTEPS, within=NonNegativeReals)	$\dot{H}_{L,FG}$ are the losses to the environment with the flue gas flow, here maximum value.
self. H_L_FG_min	var(self.GENERICCHPS, m. TIMESTEPS, within=NonNegativeReals)	$\dot{H}_{L,FG}$ are the losses to the environment with the flue gas flow, here minimum value.
self.P_woDH	var(self.GENERICCHPS, m. TIMESTEPS, within=NonNegativeReals)	P_{woDH} is the equivalent electrical power generation without district heat extraction for a constant fuel rate $\dot{H}_F$.
self.P	var(self.GENERICCHPS, m. TIMESTEPS, within=NonNegativeReals)	P is the electrical power.
self.Q	var(self.GENERICCHPS, m. TIMESTEPS, within=NonNegativeReals)	$\dot{Q}$ the heat rate extracted for district heating.
self.Y	var(self.GENERICCHPS, m. TIMESTEPS, within=Binary)	Y is the binary status variable. It equals 1, if the unit is committed or 0, when the unit is not in operation.

This is an extraction turbine in which steam is extracted during the process of heat and electricity generation. This increases the flexibility for running this turbine and thus allows for customized running according to current demand. A targeted extraction allows for perfect process integration for the turbine. In contrast to the bleeder turbine, in which steam is mainly used during the heating process, for example for preheating boiler water for food processing, an extraction turbine provides large amounts of steam for heat generation. This steam is extracted from the turbine in several places under high pressure. This allows the extraction turbine to be run based on the fluctuations of heat demand, as the household heating requirement may fluctuate considerably. If there is no demand for heating, the steam can be used for generating electricity in the low-pressure compartment of the turbine.

Extraction of steam under high pressure for heating increases the overall efficiency of the extraction power plant but reduces the overall electricity output. These different performance efficiency rates depending on the mode of use must be mapped mathematically. Thermodynamic descriptions of these modes can be found in (Nagel 2015).

If such a processing plant is to be included in an OEMOF modeled system, both operating modes must be mapped. For this a main output flow is created, but also an additional tapped out flow, representing the extraction of steam under high pressure. The main output flow passes through the turbine and is used for electricity generation. In OEMOF, a conversion for both flows is defined. Two extreme cases are to be modeled. In the full CHP mode, the maximum of steam is extracted, no electricity is produced, only heat. In the full condensing mode, only electricity is produced, and no steam is extracted. For both modes, the full CHP and the full condensing mode a conversion factor must be defined. But the variability of extraction versus electricity generation can also be defined in such a way that the pure electricity generation, i.e., full condensing mode, does not occur.

Two parameters are relevant for this class (cf. Table 2.25).

Table 2.25 Parameters of the class *ExtractionTurbineCHP* (GitHub 2017e)

Parameter	Data type	Additional information
conversion_factors	dict	Dictionary containing conversion factors for conversion of inflow to specified outflow. Keys are output bus objects. The dictionary values can either be a scalar or a sequence with length of time horizon for simulation.
conversion_factor_full_condensation	dict	The efficiency of the main flow if there is no tapped flow. Only one key is allowed. Use one of the keys of the conversion factors. The key indicates the main flow. The other output flow is the tapped flow.

Next up is the class *ExtractionTurbineCHPBlock*, in which nodes of the type *ExtractionTurbineCHP* are combined into a data block. Here a set is defined (cf. Table 2.26).

Table 2.26 Set for the class *ExtractionTurbineCHPBlock* (GitHub 2017e)

Set	Description
VARIABLE_FRACTION_TRANSFORMERS	A set with all objects of ExtractionTurbineCHP.

This class only contains one predefined parameter (cf. Table 2.27).

Table 2.27 Parameter of the class *ExtractionTurbineCHPBlock* (GitHub 2017e)

Parameter	Data type	Additional information
group	list	List of oemof.solph.Transformers (trsf) objects for which the linear relation of inputs and outputs is created, e.g., group = [trsf1, trsf2, trsf3, …]. Note that the relation is created for all existing relations of the inputs and all outputs of the transformer. The components inside the list need to hold an attribute 'conversion_factors' of type dict containing the conversion factors from inputs to outputs.

The package *oemof.solph.components* contains tools for the modeling of different components of an energy supply system.

Contents of the module constraints.py

This module is available for creating further constraints. No classes are defined in this module. One constraint refers to investments (cf. Sect. 3.3):

   def investment_limit(m, limit=None)

This constraint sets a limit for the total investment costs of an investment in an optimization function. Several parameters are required to set this constraint (cf. Table 2.28).

Table 2.28 Parameters of the constraint *investment_limit* (GitHub 2017l)

Parameter	Data type	Additional information
model	oemof.solph.model	Model to which the constraint is added.
limit	float	Absolute limit of the investment (i.e., RHS of constraint).

Another constraint is defined for emissions (cf. Sect. 3.3):

   def emission_limit(om, flows=None, limit=None)

This sets an upper limit for emissions. For calculating this, there are some key parameters (cf. Table 2.29).

Table 2.29 Parameters of the constraint *emission_limit* (GitHub 2017l)

Parameter	Data type	Additional information
om	oemof.solph. Model	Model to which constraints are added.
flows	dict	Dictionary holding the flows that should be considered in constraint. Keys are (source, target) objects of the Flow. If no dictionary is given all flows containing the 'emission' attribute will be used.
limit	numeric	Absolute emission limit.

A second constraint specifies *def equate_variables(model, var1, var2, factor1=1, name=None)*, i.e., by multiplying variable 1 one by a factor, a second variable 2 can be generated. The following parameters are used to do this Table 2.30.

Table 2.30 Parameters of the constraint *equate_variables* (GitHub 2017l)

Parameter	Data type	Additional information
var1	pyomo. environ.Var	First variable, to be set to equal with var2 and multiplied with factor1.
var2	pyomo. environ.Var	Second variable, to be set equal to (var1 * factor1).
factor1	float	Factor to define the proportion between the variables.
name	str	Optional name for the equation, e.g., in the LP file. By default the name is: equate + string representation of var1 and var2.
model	oemof.solph. Model	Model to which the constraint is added.

The example below shows how to define a transmission line in the investment mode by connecting both investment variables (GitHub 2017l):

```
>>> import pandas as pd
    >>> from oemof import solph
    >>> date_time_index = pd.date_range('1/1/2012', periods=5, freq='H')
     >>> energysystem = solph.EnergySystem(timeindex=date_time_index)
    >>> bel1 = solph.Bus(label='electricity1')
    >>> bel2 = solph.Bus(label='electricity2')
     >>> energysystem.add(bel1, bel2)
    >>> energysystem.add(solph.Transformer(
    ...    label='powerline_1_2',
    ...    inputs={bel1: solph.Flow()},
    ...    outputs={bel2: solph.Flow(
    ...        investment=solph.Investment(ep_costs=20))}))
    >>> energysystem.add(solph.Transformer(
    ...      label='powerline_2_1',
    ...      inputs={bel2: solph.Flow()},
    ...      outputs={bel1:
    ...           solph.Flow(investment=solph.Investment(ep_costs=20))}))
    >>> om = solph.Model(energysystem)
    >>> line12 = energysystem.groups['powerline_1_2']
    >>> line21 = energysystem.groups['powerline_2_1']
    >>> solph.constraints.equate_variables(
    ...      om,
    ...      om.InvestmentFlow.invest[line12, bel2],
    ...      om.InvestmentFlow.invest[line21, bel1])
```

Contents of the module custom.py

In this module user-specific components can be stored complete with constraints, block and groupings. The definitions of classes and their blocks follow one another.
  Several classes are defined:

- class *ElectricalBus(Bus)*
- class *ElectricalLine(Transformer)*
- class ElectricalLineBlock(SimpleBlock)
- class *Link(Transformer)*
- class *LinkBlock(SimpleBlock)*
- class *GenericCAES(Transformer)*
- class *GenericCAESBlock(SimpleBlock)*
- class *OffsetTransformer(Transformer)*
- class *OffsetTransformerBlock(SimpleBlock)*

The first class is called *ElectricalBus(Bus)*. The class *ElectricalBus* inherits from the class *Bus.*:

class ElectricalBus(Bus)

This is a specific electrical bus object. This bus object is used in combination with *ElectricalLine* objects. Both objects are used or linear optimal power flow (lopf) simulations.

Another class, the class *ElectricalLine* inherits from the class *Transformer*.

class ElectricalLine(Transformer)

This class is used to carry out linear optimal power flow calculations that are based on angle formulation. A bus connected to this class must be of the type *ElectricalBus*. For more information see (GitHub 2017m). Parameters for calculating this can be seen in Table 2.31.

Table 2.31 Parameter of the class *ElectricalLine* (GitHub 2017m)

Parameter	Data type	Additional information
reactance	float or array of floats	Reactance of the line to be modeled.

Next follows the class *ElectricalLineBlock*, a subclass of *SimpleBlock*.

class ElectricalLineBlock(SimpleBlock).

This is a data block for the linear relation of nodes with type class *ElectricalLine*. There is one parameter for this class (cf. Table 2.32).

Table 2.32 Parameter of the class *ElectricalLineBlock* (GitHub 2017m)

Parameter	Data type	Additional information
group	list	List of oemof.solph.Transformers (trsf) objects for which the linear relation of inputs and outputs is created, e.g., group = [eline2, trsf3, …]. The components inside the list need to hold an attribute 'reactance' of type Sequence containing the reactance of the line.

This module creates four variables, which are important for the voltage angle:

```
self.ELECTRICAL_BUSES = Set(initialize=[n for n in m.es.nodes
                                if isinstance(n, ElectricalBus)])

self.electrical_flow = Constraint(group, noruleinit=True)

self._equate_electrical_flows = Constraint(group, noruleinit=True)

self.electrical_flow_build = BuildAction(rule=_voltage_angle_relation)
```

This module also contains a class *Link(Transformer)*, which inherits from the class *Transformer*.

class Link(Transformer)

This is a link with one or two inputs and one or two outputs. There is one parameter for this class (cf. Table 2.33).

Table 2.33 Parameter of the class *Link* (GitHub 2017m)

Parameter	Data type	Additional information
conversion_factors	dict	Dictionary containing conversion factors for conversion of each flow. Keys are the connected tuples (input, output) bus objects. The dictionary values can either be a scalar or a sequence with length of time horizon for simulation.

An example for the application of this class can be found below

```
>>> from oemof import solph
    >>> bel0 = solph.Bus(label='el0')
    >>> bel1 = solph.Bus(label='el1')
    >>> link = solph.custom.Link(
    ...     label='transshipment_link',
    ...     inputs={bel0: solph.Flow(), bel1: solph.Flow()},
    ...     outputs={bel0: solph.Flow(), bel1: solph.Flow()},
    ...     conversion_factors={(bel0, bel1): 0.92, (bel1, bel0): 0.99})
    >>> print(sorted([x[1][5] for x in link.conversion_factors.items()]))
            [0.92, 0.99]
    >>> type(link)
    <class 'oemof.solph.custom.Link'>
    >>> sorted([str(i) for i in link.inputs])
            ['el0', 'el1']
    >>> link.conversion_factors[(bel0, bel1)][3]
    0.92
```

There is also a data block of the class *LinkBlock*.

class LinkBlock(SimpleBlock)

This combines relation nodes with types class under *custom.Link*. One parameter is defined for this class (cf. Table 2.34).

Table 2.34 Parameter for the class *LinkBlock* (GitHub 2017m)

Parameter	Data type	Additional information
group	list	List of oemof.solph.custom.Link objects for which the relation of inputs and outputs is created, e.g., group = [link1, link2, link3, ...]. The components inside the list need to hold an attribute 'conversion_factors' of type dict containing the conversion factors for all inputs to outputs.

Further information on this module can be found under (GitHub 2017m).

For modeling compressed air energy storage, OEMOF provides the class *GenericCAES*. This class inherits from the class *Transformer*.

class GenericCAES(network.Transformer)

The following parameters have been defined for inputs and outputs of CAES (cf. Table 2.35).

Table 2.35   Parameters for the class *GenericCAES* (GitHub 2017m)

Parameter	Data type	Additional information
electrical_input	dict	Dictionary with key-value-pair of 'oemof.Bus' and 'oemof.Flow' object for the electrical input.
fuel_input	dict	Dictionary with key-value-pair of 'oemof.Bus' and 'oemof.Flow' object for the fuel input.
electrical_output	dict	Dictionary with key-value-pair of 'oemof.Bus' and 'oemof.Flow' object for the electrical output.

An example for a modeled description of a (Compressed Air Energy Storage—CAES) can be found in (Kaldemeyer 2017).

The next class in the package *oemof.solph.components* is the class *GenericCAESBlock*. This is a data block for a node of the class *GenericCAES*. This class inherits from the parent class *SimpleBlock*.

class GenericCAESBlock(SimpleBlock)

This class combines all nodes of the type *GenericCAES* into one block. This class also contains the parameter *group* (cf. Table 2.36):

Table 2.36   Parameter for the class *GenericCAESBlock* (GitHub 2017m)

Parameter	Data type	Additional information
group	list	List containing '.GenericCAES' objects.

Several variables must be introduced for this class (cf. Table 2.37).

A new class introduced for this release of OEMOF is the *OffsetTransformer (Transformer)*. This is a specific component with exactly one input and one output. For this class, the new parameter described in Table 2.38 is introduced.

For the class listed above, there is also the block class *OffsetTransformerBlock (SimpleBlock)*. In this class, all relations of the node OffsetTransformer are combined into one block. The parameter group is also introduced for this class (cf. Table 2.39).

Table 2.37 Variables of the class *GenericCAESBlock* (GitHub 2017m)

Variable	Description
self.cmp_st	Compression: binary variable for operation status
self.cmp_p	Compression: realized capacity
self.cmp_p_max	Compression: Max. Capacity
self.cmp_q_out_s	Compression: heat flow
self.cmp_q_waste	Compression: waste heat
self.exp_st	Expansion: binary variable for operation status
self.exp_p	Expansion: realized capacity
self.exp_p_max	Expansion: Max. Capacity
self.exp_q_in_sum	Expansion: heat flow of natural gas co-firing
self.exp_q_fuel_in	Expansion: Heat flow of natural gas co-firing
self.exp_q_add_in	Expansion: heat flow of additional firing
self.cav_level	Cavern: filling levelh
self.cav_e_in	Cavern: Energy inflow
self.cav_e_out	Cavern: energy outflow
self.tes_level	TES: filling levelh
self.tes_e_in	TES: energy inflow
self.tes_e_out	TES: energy outflow
self.exp_p_spot	Spot market: positive capacity
self.cmp_p_spot	Spot market: negative capacity

TES Thermal energy storage

Table 2.38 Parameter for the class *OffsetTransformer* (GitHub 2017m)

Parameter	Data type	Additional information
coefficients	dic	Dictionary containing the first two polynomial coefficients i.e., the y-intersect and slope of a linear equation. Keys are the connected tuples (input, output) bus objects. The dictionary values can either be a scalar or a sequence with length of time horizon for simulation.

Table 2.39 Parameter for the class *OffsetTransformerBlock* (GitHub 2017m)

Parameter	Data type	Additional information
group	list	List of oemof.solph.custom.OffsetTransformer objects for which the relation of inputs and outputs is created, e.g., group = [ostf1, ostf2, ostf3, …]. The components inside the list need to hold an attribute 'coefficients' of type dict containing the conversion factors for all inputs to outputs.

Contents of the module facades.py

This module can be used to store classes that act as simplified reduced energy-specific interfaces (facades) for solph components to simplify application (GitHub 2017n).

Contents of the module groupings.py

The grouping of objects happens automatically in OEMOF, it runs as a background process. Modelers using OEMOF do not need to know anything about grouping. For this reason, the functionality of grouping objects is not of great importance for most modelers.

But for those interested in this additional information, there is a function for grouping constraints in this module:

def constraint_grouping(node)

This requires different flows to be assigned to the class *FlowWithNodes*. If you want to understand this function in depth, have a look at the source code at (GitHub 2017c).

```
investment_flow_grouping = groupings.FlowsWithNodes(
    constant_key=blocks.InvestmentFlow,
    # stf: a tuple consisting of (source, target, flow), so stf[2] is the
    # flow.
    filter=lambda stf: stf[2].investment is not None)

standard_flow_grouping = groupings.FlowsWithNodes(
    constant_key=blocks.Flow)

nonconvex_flow_grouping = groupings.FlowsWithNodes(
    constant_key=blocks.NonConvexFlow,
    filter=lambda stf: stf[2].nonconvex is not None)
```

Contents of the module models.py

This module forms a framework for an optimization model. Two classes have been predefined. The first class helps generate a basic model and is called *BaseModel*. This class inherits from the parent class *po.ConcreteModel*: This class can be found in the Pyomo package. It can be imported via:

```
import pyomo.environ as po
po.ConcreteModel
```

This class is executed in the source code by the following string:

class BaseModel(po.ConcreteModel)

The relevant parameters are listed in Table 2.40.

Table 2.40 Parameters for the class *BaseModel* (GitHub 2017f)

Parameter	Data type	Additional information
energysystem	EnergySystem object	Object that holds the nodes of an OEMOF energy system graph.
constraint_groups	list (optional)	Solph looks for these groups in the given energy system and uses them to create the constraints of the optimization problem. Defaults to :const:'Model.CONSTRAINTS'
auto_construct	Boolean	If this value is true, the set, variables, constraints, etc. are added automatically when instantiating the model. For sequential model building process set this value to False and use methods '_add_parent_block_sets', '_add_parent_block_variables', '_add_blocks', '_add_objective'.

The second class is called *Model* and inherits from the class *BaseModel:* This class is designed to create an optimization model operational and investment optimization:

class Model(BaseModel)

The following parameters are defined here (cf. Table 2.41).

Table 2.41 Parameters for the class *Model* (GitHub 2017f)

Parameter	Data type	Additional information
energysystem	EnergySystem object	Object that holds the nodes of an OEMOF energy system graph.
constraint_groups	list	Solph looks for these groups in the given energy system and uses them to create the constraints of the optimization problem. Defaults to :const:'Model.CONSTRAINTS'

The sets for this model are (cf. Table 2.42).

Table 2.42 Sets for the class *Model* (GitHub 2017f)

Set	Description
NODES	A set with all nodes of the given energy system.
TIMESTEPS	A set with all timesteps of the given time horizon.
FLOWS	A 2 dimensional set with all flows. Index: '(source, target)'.

One variable is available for this class (cf. Table 2.43).

Table 2.43 Variable for the class *Model* (GitHub 2017f)

Variable	Description
flow	Flow from source to target indexed by FLOWS, TIMESTEPS. Note: bounds of this variable are set depending on attributes of the corresponding flow object.

In some cases, additional constraints might have to be defined. Solph checks the groups of the energy system for constraint requirements and uses these to create additional constraints for the model. The reference for this is:

```
CONSTRAINT_GROUPS = [blocks.Bus, blocks.Transformer,
                     blocks.InvestmentFlow, blocks.Flow,
                     blocks.NonConvexFlow]
```

Contents of the module network.py

The classes defined in this module are required for modeling an energy system within oemof.solph. As has already been explained, an energy system is created in OEMOF as a graph or as a network of nodes. The nodes have specific constraints, which define which nodes can be connected to each other. This information is stored in the solph library. The classes are all derived from the OEMOF core network classes and have been adapted to their optimization roles. A list of classes provided in this module can serve as an overview of this module:

- class *EnergySystem(es.EnergySystem)*
- class *Flow*
- class *Bus(on.Bus)*
- class *Sink(on.Sink)*
- class *Source(on.Source)*
- class *Transformer(on.Transformer)*

As can be seen here, the energy-specific classes listed above are introduced in separate modules, where additional information about them can be stored.

The class *EnergySystem* inherits from the class *es.EnergySystem* in the solph packages. The class *es.EnergySystem* forms part of the OEMOF core. Classes in this layer are designed for modification by the solph library. This means that groupings are created as part of the modeling process.

The class *Flow* contains several parameters, which have specific characteristics. These are listed in Table 2.44.

The following example shows modeling on the basis of the solph package (see GitHub 2017). The example shows the wording of the source code in Python. This example is not given to provide a subject-specific analysis.

```
# Creating a fixed flow object. First, the letter f is
# assigned to the class 'Flow':
f = Flow(actual_value=[10, 4, 3], fixed=True, variable_costs=5)
# The parameter is called and the value is read:
f.variable_costs[1]
# The result is the integer '5'.
5
```

Table 2.44 Parameters of the class *Flow* (GitHub 2017g)

Parameter	Data type	Additional information
nominal_value	numeric	The nominal value of the flow. If this value is set the corresponding optimization variable of the flow object will be bounded by this value multiplied with min(lower bound)/max(upper bound).
min	numeric (sequence or scalar)	Normed minimum value of the flow. The flow absolute minimum will be calculated by multiplying :attr:'nominal_value' with :attr:'min'.
max	numeric (sequence or scalar)	Nominal maximum value of the flow. (see :attr:'min').
actual_value	numeric (sequence or scalar)	Specific value for the flow variable. Will be multiplied with the nominal_value to get the absolute value. If fixed attr is set to True the flow variable will be fixed to actual_value * :attr:'nominal_value', I.e., this value is set exogenous.
positive_gradient	numeric (sequence or scalar)	The normed maximal positive difference (flow $[t-1]$ < flow$[t]$) of two consecutive flow values. This means that a positive gradient is given, if the output is higher in one timestep than in the previous one.
negative_gradient	numeric (sequence or scalar)	The normed maximum negative difference (from $[t-1]$ > flow[t]) of two consecutive timesteps.
summed_max	numeric	Specific maximum value summed over all timesteps. Will be multiplied with the nominal_value to get the absolute limit.
summed_min	numeric	Specific minimum value summed over all timesteps. Will be multiplied with the nominal_value to get the absolute limit. Whenever possible, attributes are given in specific units, that are derived from an actual performance. This is particularly relevant for the investment mode. In this mode, the variable for nominal capacity can be exchanged for the investment variable.
variable_costs	numeric (sequence or scalar)	The costs associated with one unit of the flow. If this is set the costs will be added to the objective expression of the optimization problem.
fixed	boolean	Boolean value indicating if a flow is fixed during the optimization problem to its ex-ante set value. Used in combination with the:attr: 'actual_value'.

(continued)

Table 2.44 (continued)

Parameter	Data type	Additional information
investment	:class: 'oemof.solph. options.Investment' object	Object indicating if a nominal_value of the flow is determined by the optimization problem. Note: This will refer all attributes to an investment variable instead of to the nominal_value. The nominal_value should not be set (or set to None) if an investment object is used.
nonconvex		If a nonconvex flow object is added here, the flow constraints will be altered significantly as the mathematical model for the flow will be different, i.e., constraint etc. from : class:'oemof.solph.blocks.NonConvexFlow' will be used instead of :class:'oemof.solph. blocks.Flow'.

```
# Now the next parameter is called with the string 'f.actual_value'
# requiring to know the second figure in brackets:
f.actual_value[2]
# The result given is 4.
4
# Creating a flow object with time-depended lower and upper bounds:
f1 = Flow(min=[0.2, 0.3], max=0.99, nominal_value=100)
f1.max[1]
0.99
```

For the following classes *Bus(on.Bus)*, *Sink(on.Sink)*, *Source(on.Source)* no specific parameters are required. Only the class *Transformer(on.Transformer)* has a predefined parameter (cf. Table 2.45).

Table 2.45 Parameter of the class *Transformer(on.Transformer)* (GitHub 2017g)

Parameter	Data type	Additional information
conversion_factors	dict	Dictionary containing conversion factors for conversion of each flow. Keys are the connected bus objects. The dictionary values can either be a scalar or a sequence with length of time horizon for simulation.

The following example shows the relations between objects of this class (GitHub 2017g): The following example shows the modeling of the class *Transformer* (GitHub 2017g). Two aspects are relevant for this class: For the first object, there is the instruction to sort the integers for the *conversion_factors*. The instruction sorted means that items must be listed in the order of increasing value. The inputs are then

sorted alphabetically from A to Z. At the end, a new instance of the class
Transformer is defined. Inputs, outputs and the conversion factor are defined.

```
# Defining a transformer. First, the solph library is
# imported.
>>> from oemof import solph

# Next, the classes Bus and Transformer are assigned. For the
# Transformer, the inputs, outputs and their conversion factors are
# listed. At the end, the print command is given, with the following
# constraints, that the list must be sorted according to size.
>>> bgas = solph.Bus(label='natural_gas')
>>> bcoal = solph.Bus(label='hard_coal')
>>> bel = solph.Bus(label='electricity')
>>> bheat = solph.Bus(label='heat')
>>> trsf = solph.Transformer(
...      label='pp_gas_1',
...      inputs={bgas: solph.Flow(), bcoal: solph.Flow()},
...      outputs={bel: solph.Flow(), bheat: solph.Flow()},
...      conversion_factors={bel: 0.3, bheat: 0.5,
...                          bgas: 0.8, bcoal: 0.2})
>>> print(sorted([x[1][5] for x in trsf.conversion_factors.items()]))

# After the program has run, the output received is:
[0.2, 0.3, 0.5, 0.8]

# In the next step, the classes Transformer and conversion_factors are
# created for 'bel' and 'bheat' with specific values. In the final step
# the conversion_factor for bgas in timestep [3] is queried. As this is not
# given the system automatically uses the factor 1.
>>> type(trsf)
<class 'oemof.solph.network.Transformer'>
>>> sorted([str(i) for i in trsf.inputs])
['hard_coal', 'natural_gas']
>>> trsf_new = solph.Transformer(
...      label='pp_gas_2',
...      inputs={bgas: solph.Flow()},
...      outputs={bel: solph.Flow(), bheat: solph.Flow()},
...      conversion_factors={bel: 0.3, bheat: 0.5})
>>> trsf_new.conversion_factors[bgas][3]
1
```

Contents of the module options.py

This module allows to add classes to the parent class *network*. Two subclasses already created are *Investment* and *NonConvex*. For the class *Investment* the following parameters are defined in this module (cf. Table 2.46).

Table 2.46 Parameters of the class *Investment* (GitHub 2017g)

Parameter	Data type	Additional information
maximum	float	Maximum of the additional invested capacity.
minimum	float	Minimum of the additional invested capacity.
ep_costs	float	Equivalent periodical costs for the investment, if period is one year these costs are equal to the equivalent annual costs.

For the class *NonConvex* some parameters have been defined (cf. Table 2.47).

Table 2.47 Parameters of the class *NonConvex* (GitHub 2017g)

Parameter	Data type	Additional information
startup_costs	numeric	Costs associated with a start of the flow (representing a unit).
shutdown_costs	numeric	Costs associated with the shutdown of the flow (representing a unit).
minimum_uptime	numeric	Minimum time that a flow must be greater then its minimum flow after startup.
minimum_downtime	numeric	Minimum time a flow is forced to zero after shutting down.
initial_status	numeric (0 or 1)	Integer value indicating the status of the flow in the first time step (0 = off, 1 = on).

No calculations are carried out in this moduel.

2.4.5   The oemof.tools.Package

Three modules are part of this package:

– *economic.py*
– *helpers.py*
– *logger.py*

The first module, *economic.py* is used to store functions for economic calculations (GitHub 2017h). The second module *helpers.py* provides additional functions to help modelers (GitHub 2017i). This can be used for different classes. The final module, *logger.py*, contains additional functions for logging your modeling process (GitHub 2017j).

2.5 The Process of Developing Your Own Model

When you want to develop your own model for optimizing an energy supply system, OEMOF libraries and external Python libraries can be combined and customized easily. No matter how complex the requirements of your model are and how many external and internal libraries you want to integrate, each modeling process has exactly four steps (see Fig. 2.11):

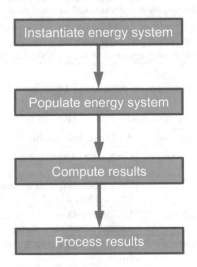

Fig. 2.11 Key steps in the modeling process (based on Hilpert 2017)

First an empty instance of an energy model is created by instantiating an existing energy system model under OEMOF. This object is the container module. This container already contains nodes and a reference to the global timeindex. Via the global timeindex, periods and steps for the optimization are defined. The container also already contains methods referring to the nodes.

In the step *populate of energy system* nodes and flows for the system are initialized based on predefined objects in OEMOF. This enables modelers to create nodes and flows in Python. In order to separate the input of data from the input of parameters, it is important to provide new functions for each new created instance of an energy system. One special feature of creating a new energy system with all its parameters in OEMOF is the fact that you can integrate CSV files (Hilpert 2017). This makes modeling easier and reduces coding errors.

Once the energy system that is to be optimized has been created in OEMOF, the optimization calculations can be executed by means of the step *compute results*. This is the moment for finding a solution based on calculations. The next step *process results* creates an output format by means of the outputlib library. Outputlib makes it easy to create diagrams and graphs and thus describes the result in various formats. The output library uses a standard data format for this.

These four steps are required for any modeling of an energy system. The following programming code shows these four steps in detail.

The following excerpt describes the instantiation of the energy system and the definition of the timeindex. Before this step, packages can be imported to provide all relevant classes, functions and methods for the optimization model.

```
import oemof.solph
# create the energy system
es = solph.EnergySystem(timeindex=date_time_index)
```

The next steps can be written in programming code in the following way: For the first two steps, assigning programming code to the steps is quite simple. Even though there are only four key steps, there are a number of additional commands in the code, that are required at this point. Sometimes, different coding systems may be used by different modelers.

```
# In order to introduce the four steps required in the program code, we use
# excerpts from the Rostock example. This is extended by
# some key components.
# The modeling is based on OEMOF objects and it is solved by means of the
# solph module. The results are displayed by means of the outputlib
# library.

# #################### Instantiate energy system ####################
# ---------------------------- Imports ----------------------------
# Default logger of OEMOF
from oemof.tools import logger
from oemof.tools import economics
import oemof.solph as solph
from oemof.outputlib import processing, views
import logging
import os
import pandas as pd
import pprint as pp

number_timesteps = 8760

# 'import pandas as pd': Pandas is a program for processing data.
# By means of the command above, Pandas is imported into the program.
# 'import logging': Logging is a Python library logging all events while
# running the program.
```

```
# 'import os': os (os=Operating System) is a standard
# library by Python [Python Tutorial 2013]. This acts as an
# interface with the operating system. This allows the
# use of functions independent of the operating system
# [PythonForBeginners 2013].

# 'from oemof.tools import logger': In the next step, the module 'looger'
# is imported from 'oemof.tools'.

# 'from oemof.tools import economics': Here, the module 'economics' is
# imported from 'oemof.tools'.

# -- Initialize the energy system and read/calculate necessary
parameters -
logger.define_logging()
logging.info('Initialize the energy system')
date_time_index = pd.date_range('1/1/2017', periods=number_timesteps,
                                freq='H')

energysystem = solph.EnergySystem(timeindex=date_time_index)

# The datetimeindex parameter is unique to Pandas. The first parameter
# sets the start date (1/1/2017). The second parameter sets
# the periods, into which the time index is to be divided.
# The third parameter sets the frequency, here set to hourly.
# More information on this can be found at [pandas o.J.-b].

# --------------------------- Read data file -------------------------
full_filename = os.path.join(os.path.dirname(__file__),
                                'storage_investment.csv')
data = pd.read_csv(full_filename, sep=',')

# this instruction requires the CSV file to be read separated by commas.

fossil_share = 0.2
consumption_total = data['demand_el'].sum()

# If the period is one year the equivalent periodical costs (epc) of an
# investment are equal to the annuity. Use OEMOF's economic tools.
epc_wind = economics.annuity(capex=1000, n=20, wacc=0.05)
epc_pv = economics.annuity(capex=1000, n=20, wacc=0.05)
epc_storage = economics.annuity(capex=1000, n=20, wacc=0.05)

# ################# Populate energy system #########################
# Next, the function 'logging.info' is called, which is to generate the
```

```
# string 'Creatig OEMOF objects' This helps check that the
# optimization model is created.
logging.info('Create OEMOF objects')
# create gas bus
r1_gas = solph.Bus(label='gas')

# create electricity bus
r1_el = solph.Bus(label='electricity')

# The following command adds the elements listed above to the
# energy system
energysystem.add(r1_gas, r1_el)

# create excess component for the electricity bus to allow overproduction
excess = solph.Sink(label='excess_r1_el', inputs={r1_el: solph.Flow()})

# create source object representing the natural gas commodity (annual
# limit)
gas_resource = solph.Source(label='rgas', outputs={r1_gas: solph.Flow(
    nominal_value=fossil_share * consumption_total / 0.58
    * number_timesteps / 8760, summed_max=1)})

# create fixed source object representing wind power plants
wt1 = solph.Source(label='wind', outputs={r1_el: solph.Flow(
    actual_value=data['wind'], fixed=True,
    investment=solph.Investment(ep_costs=epc_wind))})

# create fixed source object representing pv power plants
pv = solph.Source(label='pv', outputs={r1_el: solph.Flow(
    actual_value=data['pv'], fixed=True,
    investment=solph.Investment(ep_costs=epc_pv))})

# create simple sink object representing the electrical demand
de1 = solph.Sink(label='demand', inputs={r1_el: solph.Flow(
    actual_value=data['demand_el'], fixed=True, nominal_value=1)})

# create simple transformer object representing a gas power plant
gt1 = solph.Transformer(
    label='pp_gas',
    inputs={r1_gas: solph.Flow()},
    outputs={r1_el: solph.Flow(nominal_value=10e10, variable_costs=0)},
    conversion_factors={bel: 0.58})
```

```
# create storage object representing a battery
st1 = solph.components.GenericStorage(
    label='storage',
    inputs={r1_el: solph.Flow(variable_costs=0.0001)},
    outputs={r1_el: solph.Flow()},
    capacity_loss=0.00, initial_capacity=0,
    nominal_input_capacity_ratio=1/6,
    nominal_output_capacity_ratio=1/6,
    inflow_conversion_factor=1, outflow_conversion_factor=0.8,
    investment=solph.Investment(ep_costs=epc_storage),
)


# Again, new elements are added to the energy system.
energysystem.add(excess, gas_resource, wt1, pv, de1, gt1, st1)

# ########################### Compute results #######################
# --------------------- Optimize the energy system --------------------
logging.info('Optimize the energy system')

# Initialise the operational model.
# The class model, that inherits from the energy system,
# is called via the string 'om'
om = solph.Model(energysystem)

# If tee_switch is true solver messages will be displayed.
logging.info('Solve the optimization problem')
om.solve(solver='cbc', solve_kwargs={'tee': True})

# Next, the solver CBC is called for optimizing the energy system.
# The parameters 'tee' and 'keep' are based on Pyomo. They are not further
# explained here.
# The line 'solve_kwargs' is based on Pyomo. The string 'tee=True' is the
# command to the solver to print the output.

# ############################# Process results #######################
# ----------------------- Check and plot the results --------------
# check if the new result object is working for custom components
results = processing.results(om)

custom_storage = views.node(results, 'storage')
electricity_bus = views.node(results, 'electricity')

meta_results = processing.meta_results(om)
pp.pprint(meta_results)
```

```
my_results = electricity_bus['scalars']

# installed capacity of storage in GWh
my_results['storage_invest_GWh'] = (results[(storage, None)]
                                    ['scalars']['invest']/1e6)

# installed capacity of wind power plant in MW
my_results['wind_invest_MW'] = (results[(wind, r1_el)]
                                ['scalars']['invest']/1e3)

# installed capacity of pv power plant in MW
my_results['pv_invest_MW'] = (results[(pv, r1bel)]
                              ['scalars']['invest']/1e3)

# resulting renewable energy share
my_results['res_share'] = (1 - results[(pp_gas, r1_el)]
                           ['sequences'].sum()/results[(r1_el, de1)]
                           ['sequences'].sum())


pp.pprint(my_results)
```

References

Betz, A.: Windenergie und ihre Ausnutzung durch Windmühlen. Nachdruck der Originalausgabe aus 1926. Ökobuch Verlag, Freiburg (1994)

Durmus, M.: Die 15 wichtigsten Python Bibliotheken für Datenanalyse und maschinelles Lernen. (2017). http://www.aisoma.de/2017/11/16/die-15-wichtigsten-python-bibliotheken-fuer-datenanalyse-und-maschinelles-lernen/. Accessed 21 Nov 2017

GitHub Inc.: oemof/oemof/network.py. (2017a). https://github.com/oemof/oemof/blob/dev/oemof/network.py. Accessed 14 Dec 2017

GitHub Inc.: oemof/oemof/network.py. (2017b). https://github.com/oemof/oemof/blob/dev/oemof/energy_system.py. Accessed 14 Dec 2017

GitHub Inc.: oemof/oemof/groupings.py. (2017c). https://github.com/oemof/oemof/blob/dev/oemof/groupings.py. Accessed 18 Dec 2017

GitHub Inc.: oemof/oemof/solph/. (2017d). https://github.com/oemof/oemof/tree/dev/oemof/solph. Accessed 18 Dec 2017

GitHub Inc.: oemof/oemof/solph/components.py. (2017e). https://github.com/oemof/oemof/blob/dev/oemof/solph/components.py. Accessed 18 Dec 2017

GitHub Inc.: oemof/oemof/solph/models.py. (2017f). https://github.com/oemof/oemof/blob/dev/oemof/solph/models.py. Accessed 18 Dec 2017

GitHub Inc.: oemof/oemof/solph/network.py. (2017g). https://github.com/oemof/oemof/blob/dev/oemof/solph/network.py. Accessed 19 Dec 2017

GitHub Inc.: oemof/oemof/tools/economics.py. (2017h). https://github.com/oemof/oemof/blob/dev/oemof/tools/economics.py. Accessed 19 Dec 2017

GitHub Inc.: oemof/oemof/tools/helpers.py. (2017i). https://github.com/oemof/oemof/blob/dev/oemof/tools/helpers.py. Accessed 19 Dec 2017

GitHub Inc.: oemof/oemof/tools/logger.py. (2017j). https://github.com/oemof/oemof/blob/dev/oemof/tools/logger.py. Accessed 19 Dec 2017

GitHub Inc.: oemof/oemof/solph/blocks.py. (2017k). https://github.com/oemof/oemof/blob/dev/oemof/solph/blocks.py. Accessed 20 Dec 2017

GitHub Inc.: oemof/oemof/solph/constraints.py. (2017l). https://github.com/oemof/oemof/blob/dev/oemof/solph/constraints.py. Accessed 21 Dec 2017

GitHub Inc.: oemof/oemof/solph/custom.py. (2017m). https://github.com/oemof/oemof/blob/dev/oemof/solph/custom.py. Accessed 21 Dec 2017

GitHub Inc.: oemof/oemof/solph/facades.py. (2017n). https://github.com/oemof/oemof/blob/dev/oemof/solph/facades.py. Accessed 21 Dec 2017

GitHub Inc.: oemof organisation. (2017r). https://github.com/oemof. Accessed 21 Dec 2017

Hilpert, S. et al.: Addressing energy system modelling challenges: the contribution of the Open Energy Modelling Framework (oemof). Preprints, Basel (2017). https://pdfs.semanticscholar.org/aa20/4002418763fafe77221bc6e40778653217f2.pdf. Accessed 16 Nov 2017

Kaldemeyer, C., et al.: A generic formulation of compressed air energy storage as mixed integer linear program—unit commitment of specific technical concepts in arbitrary market environments materials today. In: Selection and/or Peer-Review Under Responsibility of 5th International Conference on Nanomaterials and Advanced Energy Storage Systems (INESS 2017). Science Direct, Elsevier, Amsterdam (2017)

Klein, B.: Python 3 Tutorial: Klassen. o.J.-a. https://www.python-kurs.eu/python3_klassen.php . Accessed 10 Nov 2017

Klein, B.: Modularisierung. o.J.-b. https://www.python-kurs.eu/modularisierung.php . Accessed 14 Nov 2017

Klein, B.: Modularisierung. o.J.-c. https://www.python-kurs.eu/python3_modularisierung.php . Accessed 15 Dec 2017

Klein, B.: Einführung in Python 3: Für Ein- und Umsteiger. 3. Auflage, Hanser Verlag, München (2018)

Maier, W.: reStructuredText (reST). The Python Wiki. (2015). https://wiki.python.org/moin/reStructuredText. Accessed 21 Feb 2018

Mollenhauer, E., et al.: Evaluation of an energy- and exergy-based generic modelling approach of combined heat and power plants. Int. J. Energy Environ. Eng. 7(2), 167–176 (2016). https://link.springer.com/content/pdf/10.1007%2Fs40095-016-0204-6.pdf. Accessed 03 Jan 2018

Nagel, J.: Nachhaltige Verfahrenstechnik. Hanser Verlag, München (2015)

oemof-developer-group: Using oemof. (2014a). http://oemof.readthedocs.io/en/stable/using_oemof.html. Accessed on 21 Nov 2017

oemof-Developer-Group: About oemof. (2014b). http://oemof.readthedocs.io/en/stable/about_oemof.html. Accessed 29 Nov 2017

oemof-developer-group: oemof-network. (2014c). Download: http://oemof.readthedocs.io/en/stable/oemof_network.html, Accessed on 29 Nov 2017

oemof-developer-group: oemof package. (2014e). http://oemof.readthedocs.io/en/stable/api/oemof.html. Accessed 05 Dec 2017

oemof-developer-group: Getting started. (2016). http://demandlib.readthedocs.io/en/latest/getting_started.html#introduction. Accessed on 21 Nov 2017

oemof-developer-group: oemof-solph. o.J.-b. http://oemof.readthedocs.io/en/latest/oemof_solph.html . Accessed 29 Jan 2018

Python Software Foundation: PyPI—the Python package index. o.J.-a. https://pypi.python.org/pypi . Accessed 21 Nov 2017

Python Software Foundation: 9. Classes. (2018a). https://docs.python.org/2/tutorial/classes.html. Accessed 13 Dec 2017

Python Software Foundation: Project Summaries. (2018b). https://packaging.python.org/key_projects/#pip. Accessed 21 Nov 2017

Python Software Foundation: 6. Modules. (2018c). https://docs.python.org/3/tutorial/modules.html . Accessed on 06 Jan 2018

Python Software Foundation: 8.4. collections.abc—Abstract Base Classes for Containers. (2018f). https://docs.python.org/3/library/collections.abc.html. Accessed 19 Jan 2018

pandas: MultiIndex/Advanced Indexing. o.J.-a. http://pandas.pydata.org/pandas-docs/stable/advanced.html . Accessed 05 Dec 2017

pandas: pandas.date_range. (2017b). http://pandas.pydata.org/pandas-docs/stable/generated/pandas.date_range.html, Accessed on 07 Dec 2017

Quaschning, V.: Regenerative Energiesysteme: Technologie – Berechnung - Simulation, 9th edn. Hanser Verlag, München (2015)

Ratka, A. et al, Ehrmeier, B. (Ed.): Technik Erneuerbarer Energien. UTB Band-Nr. 4343, Verlag Eugen Ulmer KG, Stuttgart, 2015

ZetCode: Python keywords. o.J. http://zetcode.com/lang/python/keywords/ . Accessed 13 Nov 2017

Chapter 3
Mathematical Description of the Objects

The OEMOF system of programs with its object-oriented approach offers several classes designed to model an energy supply system of a specified region, in a specified location or based on a specific set of constraints. Understanding the mathematical conceptualization of the different technical components but also understanding the economic conceptualization helps modelers to gain a deeper knowledge of their own energy model.

When looking at the mathematics behind the energy supply model, there are two key factors. The first factor is the objective function. This objective function is a very complex equation. Depending on the question that is to be answered and the objective of the modeling, this mathematical equation can be composed of technical, ecological and economic terminology. The other factors of complexity are the constraints. Constraints can be used to ensure the compliance with emission limits or to meet predefined optimal running conditions. These types of constraints are often described by inequations, i.e., equations requiring something to stay below or above a set limit.

Translating the mathematical conditions into programming code can be carried out in different ways. For this reason, the implementations shown here may experience several redevelopments. But once a general understanding of how to translate the requirements for an energy system into equations or inequations has been arrived at, it will be easier to model your own requirements of an energy system using OEMOF.

In the following section, we will look at mathematical formulae, which either formulate part of the objective function or define constraints. We will present the mathematical configuration of the different classes already introduced in Chap. 2:

- Bus
- Sink
- Source

© Springer International Publishing AG, part of Springer Nature 2019
J. Nagel, *Optimization of Energy Supply Systems*, Lecture Notes in Energy 69,
https://doi.org/10.1007/978-3-319-96355-6_3

– Transformer
– Flow

The mathematical description of these classes will be presented in the order of the modules.

As has already been described, OEMOF presents an energy model as a bipartite graph with edges and nodes, in which components are always linked by buses and buses are only ever linked to components.

The nodes in the bipartite graph represent entities, which are described by the base class *Entity* (oemof 2014b). This base class *Entity* is mainly relevant for the developer. Modelers will not really work with this class.

An entity can either represent a bus or a component. Entities can be divided into the following classes or subclasses:

– *Entity*

  – *Bus*
  – *Component*

    – *Sink*

      – *Simple*

    – *Source*

      – *Commodity*
      – *DispatchSource*
      – *FixedSource*

    – *Transformer*

      – *Simple*
      – *CHP*
      – *SimpleExtractionCHP*
      – *Storage*
      – *ElectricalLine**
      – *Link**
      – *GenericCAES**
      – *OffsetTransformer**

    – *Transport*

      – *Simple*

The subclasses with an asterisk are located in the module *custom.py* and represent specific classes. They were created by users and were included into the OEMOF system of programs after having been thoroughly tested and applied by OEMOF developers.

The approach described here can be derived from the Entity Relationship Modeling (ERM) approach. In an ERM approach, a system is modeled by making use of (Chen 1976):

- Objects (Entities),
- their attributes and
- their relationships.

In OEMOF, the relationships between objects are described by flows. The attributes are assigned to objects via the programming code.

Graph theory is helpful for understanding the overall setup of modeling in OEMOF. Graph theory is also used in computer science or in network planning. In graph theory, each graph G is composed of a set of nodes V (these are called *vertices* in graph theory; but in OEMOF we use the term *nodes*) and of a set of edges, these are labeled K (using the German word for edge, *Kante*). As mentioned earlier, the edges represent the relationships, also called relations, between nodes. A simple graph G would be a simple, non-directional graph without double edges or slings. The set K is a subset of the bipartite subset of V. This means that each edge is a set of two nodes. The nodes do not have a set order and do not have a specific direction. This leads to the following mathematical mapping: see Eq. 3.1 (Werners 2013):

$$G = (V, K) \tag{3.1}$$

The graphic representation is shown in Fig. 3.1.

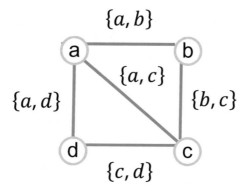

Fig. 3.1 Example of a simplified Graph $G = (V, K)$ where $V = \{a, b, c, d\}$ and $K = \{\{a, b\}, \{a, c\}, \{a, d\}, \{b, c\}, \{c, d\}\}$ (based on Domschke 2015)

It can be stated that each edge $k \in K$ links two separate nodes (vertices) at the two ends of its edge. Thus, each edge is a bipartite set with the nodes u and v, $u \in V$ and $v \in V - \{u\}$ as its elements. Each edge consists of a pair of nodes $\{u, v\}$. Equations 3.2 and 3.3 describe these relations:

$$k = \{u, v\} \tag{3.2}$$

$$K \subseteq \{\{u, v\} | u \in V, v \in V - \{u\}\} \tag{3.3}$$

In set theory, the symbol $\subseteq$ means 'is a subset of'. The symbol $\in$ means 'is an element of'. In the following section, capital letters represent sets and lower case letters elements in that set, as in the example $V = (u_1, u_2, u_3, \ldots)$ a set comprising several instances of u_i. If the graph has a direction, normal brackets () are used instead of curly brackets { }.

This then shows that K is a subset of the power set of V (set of nodes − vertices) (cf. Eq. 3.4) (Domschke 2015):

$$K \subseteq p(V) \text{ or } K \subseteq VxV \tag{3.4}$$

The term VxV represents the cross product of the set V.

The simpler formula given in Fig. 3.1 of the graph $G = (V, K)$ is now shown to be composed of the nodes (vertices) V with their elements a, b, c, d (cf. Eq. 3.5):

$$V = \{a, b, c, d\} \tag{3.5}$$

together with the set of edges K composed of the tuples (pairs) of nodes (vertices) (cf. Eq. 3.6):

$$K = \{\{a, b\}, \{a, c\}, \{a, d\}, \{b, c\}, \{c, d\}\} \tag{3.6}$$

As models in OEMOF are normally represented as directed graphs, we will have a look at directed graphs next. In a directed graph nodes are connected by arrows, with one node at the start point and one node at the end point. But still V represents the set of nodes with its elements a, b, c and d. In the same way edges K are composed of a tuple of vertices as represented in the following equation (cf: Eq. 3.7):

$$\vec{K} = \{(a, b), (a, c), (a, d), (b, c)\} \tag{3.7}$$

To indicate that this is a directed graph, the tuples are described as the set $\vec{K}$ with an arrow symbol above the set letter. But Eq. 3.4 remains valid.

OEMOF is based on a bipartite graph, which is a unique type of graph. The set of nodes in a bipartite graph are two disjoint subsets. This means that the set of K is divided into two subsets A and B which do not share a single element. One example of two disjoint sets would be the set of buses and the set of components. These represent two very different types of nodes. Edges in OEMOF always connect nodes from separate subsets, i.e., they connect two different types of nodes. An example of this can be seen in Fig. 3.2.

Fig. 3.2 Example of a bipartite graph

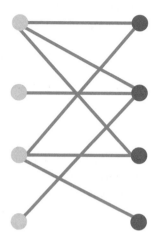

In order to demonstrate which component, e.g., CHP plant, supplies what amount of energy via a bus to another component, e.g., a private household requiring heat, this can be modeled by means of a bipartite graph. In such a model, the energy supply would always be represented by an edge which would always link a component with a bus. In graph theory, this can be conveyed by the following equation (cf. Eq. 3.8):

$$G = (V_1, V_2, K) \text{ with } V_1 \cap V_2 = \{\} \text{ for } i \neq j \text{ and } K \subseteq \{\{u, v\} | u \in V_1, v \in V_2\} \quad (3.8)$$

The vertical line | in Eq. 3.8 corresponds to and is read aloud as the phrase: 'where [u] is an element of'. Thus this equation represents the following information: Set K is a subset of the tuple $\{u, v\}$ where u is an element of subset V_1 and v is an element of subset V_2.

Going back to the OEMOF system of programs we can apply our knowledge of graph theory to modeling energy systems in OEMOF. The constraints described here are part of the mathematical model that will be introduced in more detail in the following section. By means of graph theory, constraints can be described with a limited set of symbols (see oemof 2014b).

In OEMOF, the vertices are called *nodes* or *Entities*. In order to continue with this convention, the nodes are represented by the letter E for entity. The edges K of an energy system are energy flows. These are the inputs and outputs of the entities. These are always directed graphs, as each flow has a start point and an end point, thus describing a direction. In the following equations, inputs are represented by the letter I and elements of this set are represented by a lower case i. The same is true for outputs. The set of outputs is O and elements of this set are represented by o.

Inputs and outputs are represented as edges which connect nodes to each other. In the completed graph, each edge has a value (Domschke 2015). This value is called the weight of the edge. Values can be costs, e.g., € per kWh or emissions in grams per kWh.

In OEMOF the set of nodes can be represented as a set of entities E and their elements are represented by e. Set E and its elements e are a unified set of the sets of buses (B), the set of transformers (T), the set of sources (S), the set of sinks (K) and the set of transports (P) as shown in Eq. 3.9:

$$E := E_B \cup E_T \cup E_S \cup E_K \cup E_P \tag{3.9}$$

A set of components E_C can be described in the following formula (cf. Eq. 3.10):

$$E_C := E \backslash E_B \tag{3.10}$$

Equation 3.10 in set theory conveys the following information: The set of components E_C represents the entire set of entities E minus the set of buses E_B.

Each edge links two entities (e_m, e_n). As a result in OEMOF an edge represents the input of one entity (e_n) plus the output of another neighboring entity (e_m). The edge thus has a direction. A set of directed edges $(\vec{E})$ thus represents a pair of entities, connected by a directed edge, as shown in Eq. 3.11.

$$\vec{E} := \{(e_m, e_n), \dots\} \tag{3.11}$$

where

$m, n \in \mathbb{N}^*$ Number of the entity, number must be element of the set of positive integers.

Each pair is given in a specific order, which is expressed via the letters assigned to the elements in brackets (e_m, e_n), where e_m corresponds to the start point and e_n to the end point.

Within the OEMOF graph model, there are never any undirected edges. To explain this, we will look at two examples. First let us look at an electricity storage plant. This will be filled within one period of time and emptied within another period of time. This means that the direction of the flow changes. In order to reflect this in OEMOF, the component energy storage plant is connected to a bus via two parallel edges with different directions. These double edges were not explicitly pre-modeled in the Release v0.1.0. Therefore in Release v0.1.0 electricity storage units could be filled and emptied within the same period of time, which is of course impossible in the real world. This caused incorrect results in the modeling. The exact method for modeling double directional edges in Release v0.2 has not been fully described yet and will have to wait for the release date. Another problematic aspect for flow direction is transformers. Transformers can also be cables. This means that transformer cables physically allow for a two-directional flow. In OEMOF Release v0.1.0 this was carried out by modeling two separate objects; one object was the 'transformer part' providing flow in one direction, the other object was the 'cable part' for the flow in the other direction. We will have to find out how this problem will have been resolved in Release v0.2.0

The connections described by the equations above are described in Fig. 3.3.

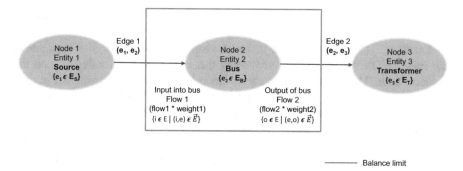

Fig. 3.3 General overview of nodes and directed edges according to Eqs. 3.9 to 3.11

Each node is the start point or end point of a directed edge. This means that there is at least one inflow (i) or outflow (o). These inflows and outflows can be described as subsets I_e and O_e of the directed edge $\vec{E}$, each subset being assigned to an entity e (cf. Eqs. 3.12 and 3.13). Inflows are assigned to the entities they flow towards e_n, while outflows are assigned to entities e_m, which they flow from.

$$I_e := \{i \in E | (i, e) \in \vec{E}\} \tag{3.12}$$

$$O_e := \{o \in E | (e, o) \in \vec{E}\} \tag{3.13}$$

Now each entity must have a clear balance delimitation. Each balance limit must be described by certain constraints. Constraints always apply to the entities of the components (E_C). For these to be calculated, function f is described as the 'transfer function' for each component (cf. Eq. 3.14):

$$f(I_e, O_e) \leq \vec{0}, \quad \forall e \in E_C \tag{3.14}$$

Equation 3.14 is an abstract formula reflecting the fact that for each entity e comprising the set of entities of the components (E_C) there is always a certain loss during the transformation of inputs to outputs. If you calculate the balance of a component, the difference between output and input is always lower than or equal to zero, this is due to the transformation losses. But if we look at an example taken from real life, e.g., a CHP plant, the energy input does not correspond to the energy output, due to transformation losses. This aspect is reflected in Eq. 3.14. In the model, transformation losses are always attributed to the component. The symbol $\forall$ means 'all' in mathematics.

This equation indicates that each output (o_e) must always correspond to the input for the next entity (i_e). For this reason, the statement still applies that the balance of an edge between two entities (e_m, e_n) must always be zero (cf. Eq. 3.15):

$$o_e - i_e = 0 \qquad (3.15)$$

Equation 3.15 reflects the fact, that each output of an entity (o_e) corresponds to the input (i_e) of the following entity. How this is defined exactly depends on the balance limit. This is the case if the balance limit is set as shown in Fig. 3.3. Whereas, if the balance limit delimitation is based around entity 1, edge 1 will not correspond to an input but will be an output. The balance of an edge must always be zero, irrespective of the balance delimitation.

Based on this general understanding of edges, entities, inflows, outflows and balance limits the next chapter will introduce the constraints for certain modules. In order to be able to engage separately with each module, each module will be accessed via the OEMOF organization page. (cf. Fig. 2.7). The second and third layer can be accessed via the OEMOF button (cf. Figs. 3.4 and 3.5).

Access to the fourth layer, which is relevant for this section, is by means of a link.

Fig. 3.4 Second layer of the OEMOF platform (GitHub 2018b)

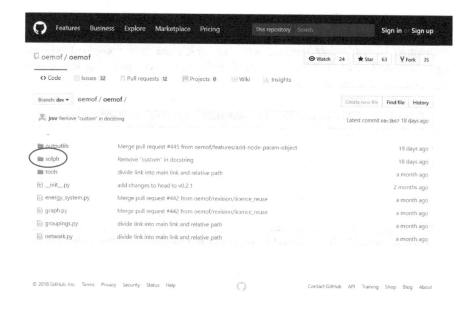

Fig. 3.5 Third Layer of the OEMOF platform (GitHub 2018a)

Now the modules can be described in more detail, as shown in Fig. 3.6.

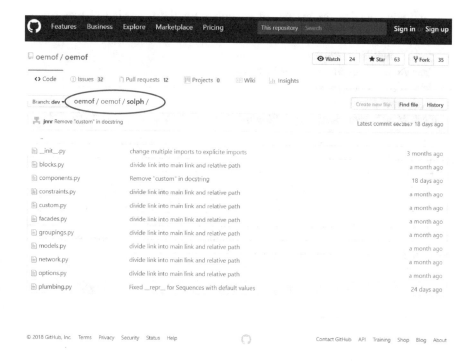

Fig. 3.6 Fourth layer of the OEMOF platform (GitHub 2017a)

The modeling of mathematical formulae is carried out in the package oemof.-solph. The mathematical formulae provided in this package are inequations for the formulation of constraints, several terms of which are then combined into one objective function. Further methods and functions are given to define the rules of calculating parameters or variables.

The OEMOF programming code provides functions in the form of inner or nested functions. This means that the functions can only be used or seen in the context of the surrounding method or function. In the following sections, inner or nested functions are simply referred to as functions. The surrounding method is normally encoded in the following way in OEMOF:

```
def _create(self, group=None):
```

Parameters, sets, variables, and constraints are all created by means of this surrounding method. This method is normally not described in the following sections, as this was already introduced in Chap. 2.

The excerpts of source code given below contain comments to help describe certain features within the programming code.

3.1 Mathematical Equations in the Module blocks.py

This module contains the system of equations for the following classes (oemof 2014a):

- class *Flow(SimpleBlock)*
- class *InvestmentFlow(SimpleBlock)*
- class *Bus(SimpleBlock)*
- class *Transformer(SimpleBlock)*
- class *NonConvexFlow(SimpleBlock)*

All variables, parameters, and sets created in the following equations and inequations have been introduced in Chap. 2. In the following sections, these are briefly described again to help with the overall conceptualization.

All classes listed here are child classes of the parent class *SimpleBlock* in Pyomo. All subclasses of *SimpleBlock* are methods to print block information and display values in the block. The class *SimpleBlock* itself inherits from the Pyomo class *Block*. These classes are all indexed components that contain other components (Pyomo 2018). These may or may not be blocks as well.

3.1.1 Class Flow(SimpleBlock)

The energy input and output of a component are described by the class *flow*. A flow corresponds to an edge in graph theory. An energy flow may include ecological factors, such as emissions, or may have economic factors, such as investment costs. These can be formulated in the equations in the form of constraints (see oemof 2014c).

The first constraint is that a flow must always be a positive number or be zero. Let us look at a case where an energy storage unit does not only store energy, but also feeds energy back to the grid. This must be modeled via a double edge describing two separate flows. The flow in the one direction represents the input, the flow in the other direction the output. For each flow in each direction the statement is still valid that the balance must be either positive or zero. In order to ensure that there is no excess energy, it is a good idea to create a *sink* for the *excess*. To ensure that there is no shortage, it is always a good idea to introduce a component *source* to represent potential shortage.

A second constraint is to set a maximum flow for each timestep. This maximum does not have a specific dimension. After multiplying this maximum value with a nominal value, an output maximum can be calculated. This means that a flow is assigned a maximum load that must not be exceeded.

The following third constraint describes a maximum for the total sum of a flow during a set time period t. The value of the variable $flow(i, o, t)$ is added up across all (timesteps) t (Eq. 3.16). The product of all timesteps for the variable $flow(i, o, t)$ is an absolute number without dimension. This number must not exceed a *summed_max*. The parameter *summed_max* is also a set number without unit dimension. The maximum value is multiplied by nominal value (*nominal_value*, data type: numeric), which is the basis for calculating the maximum energy provided by this flow.

$$\sum_t flow(i, o, t) \cdot \tau \le summed\_max(i, o) \tag{3.16}$$

In this equation:

τ is a timestep, consisting of 1-h value $\tau = 1$, and ½-h $\tau = 0, 5$
This equation applies to:

$$\forall(i, o) \in SUMMED\_MAX\_FLOW$$

The variable $flow(i, o, t)$ is based on the class *Model* (cf. Table 2.43). The expression *SUMMED_MAX_FLOW* represents a subset of all flows, for which the attribute *summed_max* is *not None*.

As an example for such a maximum value, see the following code excerpt from *basic_simple.py*:

```
# Create source object representing the natural gas commodity 'rgas'
# (annual limit). This object has the output flow 'bgas'. 'bgas' is the bus
# for natural gas.
energysystem.add(solph.Source(label='rgas', outputs={bgas: solph.Flow(
          nominal_value=29825293, summed_max=1)}))
```

As an example of this, the formula is written in reStructuredText (reST) in the following way:

```
Flow max sum :attr:'om.Flow.summed_max[i, o]'
   .. math::
     \sum_t flow(i, o, t) \cdot \tau \leq summed\_max(i, o), \\
     \forall (i, o) \in \textrm{SUMMED\_MAX\_FLOWS}
```

The expression *om* written before the class *Flow* (*om.Flow.summed_max...*) stands for *OperationsModel*. In newer Releases from v0.2.0 this is simply written as *model*. This is a typical least-cost balancing model. This is a fully created cost-calculating model, which is provided to modelers in OEMOF. But modelers also have the option to create their own models by modifying the existing least-cost balancing model.

Just as there are maximum values for flows, so there are minimum values. This is described in the next constraint. This sets a minimum for the sum of all variables for flow(i, o, t) for all timesteps (cf. Eq. 3.17):

$$\sum_t flow(i, o, t) \cdot \tau \geq summed\_min(i, o) \qquad (3.17)$$

This applies to:

$$\forall (i, o) \in SUMMED\_MIN\_FLOWS$$

Limiting flows within a time period or across all time periods is only one aspect of modeling reality. It may also be relevant to describe changes in flow from one timestep to the next one. This might also require modelers to set maximum and minimum values when running power plants. In order to understand this more clearly we will look at two different systems for running power stations. For this, we will always look at two consecutive timesteps of a flow.

One option to take power stations into and out off the grid is to switch plants on or off. One example for this is running a biogas power plant with several CHP units. Depending on demand, CHPs can be fed into the grid, switched on or off. CHPs run big generators that can be switched on or off quite easily, so that the switching on or off can take less than one timestep. If a CHP unit is added, this reaches its maximum load very quickly, e.g., 75 kW. This means that the flow change from one timestep to the next corresponds exactly to that output. If a CHP is switched off, the output is reduced exactly by that same amount. Thus the amount of change is exactly the same, only in a negative or a positive direction. This indicates that changes will often take place in set increments. In order to reflect this in the model, the absolute minimum value of change must be set.

The second way of running an energy system is changing the load profile of one plant. This may be relevant to coal power stations, as their energy output cannot be changed so easily. Load changes happen during the switching on and off process. In order to ensure that the load profile does not change too drastically, as would be the case during the switching on or off of one power station, a maximum value for the change value must be set. In this case, a maximum for the absolute change value of a flow must be set.

In order to describe in OEMOF how changes in operation levels of power plants can take place, maximum and minimum change levels for flows must be set. To make this work, two parameters are introduced, the positive and the negative gradient. The positive gradient represents an upward change, for example a switching on of a power plant, the negative gradient corresponds to a downward change in energy level. Both parameters require a maximum and a minimum level. This means that there are four limit values in total. Let us first have a look at the positive gradient. Two different kinds of limits can be set here:

1. For each time period only a stepwise change may take place. Power plants can only be added or subtracted in certain increments, i.e., plants can only be switched to by increments. This requires setting a minimum increment value for load changes between two timesteps. The change of a flow must therefore be above a set minimum increment.
2. The speed of a load change must not exceed a certain maximum value. This predefined maximum (maximum speed for load change) ensures that load changes can only happen in set increments. This means that the load alteration of a flow has an upper limit.

To summarize: While the positive gradient looks at adding power stations (case 1) or switching on a power station (case 2), the negative gradient refers to taking off power stations (reverse case 1) i.e., powering down a plant (reverse case 2). In OEMOF Release v0.2.0 the following two new equations (Eqs. 3.18 and 3.19) have been created to formulate a constraint for case one, but both for the positive and the negative gradient.

For case 1, the constraint for the positive gradient can be formulated in the following term (Eq. 3.18):

$$flow(i,o,t) - flow(i,o,t-1) \geq positive\_gradient(i,o,t) \qquad (3.18)$$

This applies to:

$$\forall(i,o) \in POSITIVE\_GRADIENT\_FLOWS \text{ and } \forall t \in TIMESTEPS$$

The constraint for the negative gradient is described by Eq. 3.19:

$$flow(i,o,t-1) - flow(i,o,t) \geq negative\_gradient(i,o,t) \qquad (3.19)$$

This applies to:

$$\forall(i,o) \in NEGATIVE\_GRADIENT\_FLOWS \text{ and } \forall t \in TIMESTEPS$$

The values given in Eqs. 3.18 and 3.19 are numeric and do not have a specified dimension. The terms *POSITIVE_GRADIENT_FLOWS* and *NEGATIVE_ GRADIENT_FLOWS* indicate that these are subsets of flows for which the attribute positive_gradient or negative_gradient is set to not None.

In the module these constraints are again written in reST, here is the formulation for the negative gradient:

```
Negative gradient constraint :attr:'om.Flow.negative_gradient_constr
[i, o]':
    .. math:: flow(i, o, t-1) - flow(i, o, t) \geq \
    negative\_gradient(i, o, t), \\
    \forall (i, o) \in \textrm{NEGATIVE\_GRADIENT\_FLOWS}, \\
    \forall t \in \textrm{TIMESTEPS}
```

The description in reST is mainly intended for your documentation. The reST version can be converted directly to readable text, without any further processing. These reSt versions are not shown in the rest of this chapter.

This class is used to write part of the objective function (cf. Eq. 3.20). But this function can only be used if the model contains the parameter 'variable_costs'.

$$\sum_{(i,o)} \sum_t flow(i,o,t) \cdot variable\_costs(i,o,t) \qquad (3.20)$$

NB: Please note that sums calculated for the flow will always either calculate the sum for i,o, as is the case in Eq. 3.20, then the sum is always either calculated for i or for o.

It is also important to note that for both, the positive and the negative gradient, the value is always multiplied by the *nominal_value*:

```
# Inputs (i), outputs (o) and flows (f) are grouped
# together, so that the equations do not have to be
# formulated separately for each element: 'for i, o, f in group:'.


# In the following term: 'm.flow[i, o]': 'm' refers to the parent
# object. This is defined further on in the code by:
# 'm = self.parent_block()'
# [oemof 2014d]. The function 'parent_block()' is based on
# Pyomo. This function returns the parent of this object.
```

```
# The term 'positive_gradient' is the attribute of a flow. This
# is a dictionary. The attribute has two parameters.
# The first parameter 'ub' is the upper limit (upper bound) of the flow
# and the second parameter is the timestep, for which the value '0'
# has been set. The timestep t = 0 is the first timestep.
# If the gradient for the first timestep is 'None', it follows
# that all future timesteps also have the value 'None'
# which is why this calculation is not reiterated. Thus only the first
# timestep is read. If the value is 'not None', the iteration continues
# across the next timesteps.
# The term 'setub' refers to: 'set upper bound'
for i, o, f in group:
    if m.flows[i, o].positive_gradient['ub'][0] is not None:
        for t in m.TIMESTEPS:
            self.positive_gradient[i, o, t].setub(
                f.positive_gradient['ub'][t] * f.nominal_value)
    if m.flows[i, o].negative_gradient['ub'][0] is not None:
        for t in m.TIMESTEPS:
            self.negative_gradient[i, o, t].setub(
                f.negative_gradient['ub'][t] * f.nominal_value)
```

A group consists of a list of tuples for the separate elements of a group. This list contains the flow object *f*, the associated source *s* and the target *t*, e.g., a storage unit. For a group, the term would look like this in this class (cf. Eq. 3.21) (oemof 2014d):

$$groups = [(s1, t1, f1), (s2, t2, f2), \ldots] \tag{3.21}$$

Equations 3.16–3.20 of this class are implemented in several different functions.
The first two functions create the two constraints for the calculation of the maximum and the minimum bounds for the sum of flows in this block. These generate the values for the set *self.SUMMED_MAX_FLOWS*. For the upper bound (maximum sum) the following term is created (GitHub 2017d):

```
def _flow_summed_max_rule(model):
    for inp, out in self.SUMMED_MAX_FLOWS:
        lhs = sum(m.flow[inp, out, ts] * m.timeincrement[ts]
            for ts in m.TIMESTEPS)
        rhs = (m.flows[inp, out].summed_max *
            m.flows[inp, out].nominal_value)
        self.summed_max.add((inp, out), lhs <= rhs)
self.summed_max = Constraint(self.SUMMED_MAX_FLOWS, noruleinit=True)
self.summed_max_build = BuildAction(rule=_flow_summed_max_rule)
# The penultimate line limits the sum ('summed_max') for a flow
# via the Pyomo function 'Constraint' [Pyomo 2014].
# The next line uses a BuildAction function from Pyomo.
```

```
# This function triggers actions to be done as part of the model building
# process [Pyomo 2014]. In this case, the listed function is to
# be carried out.
```

For the lower bound (minimum) the next function is implemented _flow_
summed_min_rule(model):

```
def _flow_summed_min_rule(model):
    for inp, out in self.SUMMED_MIN_FLOWS:
        lhs = sum(m.flow[inp, out, ts] * m.timeincrement[ts]
            for ts in m.TIMESTEPS)
        rhs = (m.flows[inp, out].summed_min *
            m.flows[inp, out].nominal_value)
        self.summed_min.add((inp, out), lhs >= rhs)
self.summed_min = Constraint(self.SUMMED_MIN_FLOWS, noruleinit=True)
self.summed_min_build = BuildAction(rule=_flow_summed_min_rule)
```

In order to calculate the positive and the negative gradient two more functions
are required. The function for the positive gradient is _positive_gradient_flow_rule
(model):

```
def _positive_gradient_flow_rule(model):
    for inp, out in self.POSITIVE_GRADIENT_FLOWS:
        for ts in m.TIMESTEPS:
            if ts > 0:
                lhs = m.flow[inp, out, ts] - m.flow[inp, out, ts-1]
                rhs = self.positive_gradient[inp, out, ts]
                self.positive_gradient_constr.add((inp, out, ts),
                    lhs <= rhs)
            else:
                pass # return(Constraint.Skip)
self.positive_gradient_constr = Constraint(
    self.POSITIVE_GRADIENT_FLOWS, noruleinit=True)
self.positive_gradient_build = BuildAction(
    rule=_positive_gradient_flow_rule)
```

This is the constraint for the negative gradient of the flow: _negative_gradi-
ent_flow_rule(model):

```
def _negative_gradient_flow_rule(model):
    for inp, out in self.NEGATIVE_GRADIENT_FLOWS:
        for ts in m.TIMESTEPS:
            if ts > 0:
                lhs = m.flow[inp, out, ts-1] - m.flow[inp, out, ts]
```

```
            rhs = self.negative_gradient[inp, out, ts]
            self.negative_gradient_constr.add((inp, out, ts),
                                  lhs <= rhs)
        else:
            pass # return(Constraint.Skip)
self.negative_gradient_constr = Constraint(
    self.NEGATIVE_GRADIENT_FLOWS, noruleinit=True)
self.negative_gradient_build = BuildAction(
    rule=_negative_gradient_flow_rule)
```

If the flow is measured in integers, the function _integer_flow_rule(block, i, o, t) is used. By means of this function, the variables of the flow are defined as non-negative integers:

```
def _integer_flow_rule(block, i, o, t):
    return self.integer_flow[i, o, t] == m.flow[i, o, t]

self.integer_flow_constr  =  Constraint(self.INTEGER_FLOWS,  m.TIMESTEPS,
                        rule=_integer_flow_rule)
```

Now the method can be used for calculating the values of the objective function according to Eq. 3.20. This function applies to all flows that have one fixed-cost and one variable-cost part:

```
def _objective_expression(self):
    m = self.parent_block()

    variable_costs = 0
    gradient_costs = 0

    for i, o in m.FLOWS:
        if m.flows[i, o].variable_costs[0] is not None:
            for t in m.TIMESTEPS:
# The following sentence is a shorthand which means:
# A += B corresponds to: A = A + B.
                variable_costs += (m.flow[i, o, t] * m.timeincrement[t] *
                        m.flows[i, o].variable_costs[t])

        if m.flows[i, o].positive_gradient['ub'][0] is not None:
            for t in m.TIMESTEPS:
                gradient_costs += (self.positive_gradient[i, o, t] *
                        m.flows[i, o].positive_gradient['costs'])
```

```
if m.flows[i, o].negative_gradient['ub'][0] is not None:
    for t in m.TIMESTEPS:
        gradient_costs += (self.negative_gradient[i, o, t] *
                m.flows[i, o].negative_gradient['costs'])

    return variable_costs + gradient_costs
```

3.1.2   Class InvestmentFlow(SimpleBlock)

When modeling an optimization function for the energy supply in a certain region, the optimization aspects tend to be the costs. This is the reason why flows in OEMOF can be assigned to costs. For the objective function to be calculated constraints must be set. One such constraint computes investment costs for building a power plant (fixed investment) (oemof o.J.-c.). For this an actual value is entered as the flow variable. In order to receive the absolute value of the flow the actual value is multiplied by the nominal value. In order to enter investment costs as a flow the actual value of an input or output is multiplied by the investment costs per timestep (cf. Eq. 3.22):

$$flow(i, o, t) = actual\_value(i, o, t) \cdot invest(i, o) \tag{3.22}$$

This applies to:

$$\forall (i, o) \in FIXED\_FLOWS \text{ and } \forall t \in TIMESTEPS$$

The factor *invest* is a variable which reflects costs. The parameter *actual_value* reflects the desired value (actual value) and is numeric.

The term *FIXED_FLOWS* shows that this is a subset of flows for which the attribute 'fixed' is set to '*true*'. The parameter *fixed* means that this flow is fixed at a set value while optimization is running (cf. Table 2.44). Subsequently, these add-ons which can be sets or subsets and to which each equation applies have to be read in this way. The term will always state which attribute must be set to which value in order for the term to apply.

For these equations, lower bounds for cost flows are normally of vital importance (cf. Eq. 3.23):

$$flow(i, o, t) \geq \min(i, o, t) \cdot invest(i, o) \tag{3.23}$$

This applies to:

$$\forall(i,o) \in \mathit{MIN\_FLOWS} \text{ and } \forall t \in \mathit{TIMESTEPS}$$

Equation 3.22 sets a minimum value for a flow for each timestep $\min(i,o,t)$. The minimum again is a numeric value. The subset $\mathit{MIN\_FLOWS}$ contains all flows, for which the value for the first timestep is set to *not None*. The upper bound for the invest flows works in a similar way to the minimum value as shown in Eq. 3.24:

$$flow(i,o,t) \leq \max(i,o,t) \cdot invest(i,o) \tag{3.24}$$

This applies to:

$$\forall(i,o) \in \mathrm{MAX\_FLOWS} \text{ and } \forall t \in \mathit{TIMESTEPS}$$

Another constraint must define the specific maximum value summed over all timesteps of the flow (cf. Eq. 3.25):

$$\sum_t flow(i,o,t) \cdot \tau \leq summed\_\max(i,o) \cdot invest(i,o) \tag{3.25}$$

This applies to:

$$\forall(i,o) \in \mathit{SUMMED\_MAX\_FLOWS}$$

There is also a constraint for the specific minimum value summed over all timesteps of the flow (cf. Eq. 3.26):

$$\sum_t flow(i,o,t) \cdot \tau \geq summed\_\min(i,o) \cdot invest(i,o) \tag{3.26}$$

This applies to:

$$\forall(i,o) \in \mathit{SUMMED\_MIN\_FLOWS}$$

Now the objective function can be created. This is the term for the equivalent periodical costs (epc) (cf. Eq. 3.27):

$$\sum_{i,o} invest(i,o) \cdot ep\_costs(i,o) \tag{3.27}$$

The parameter *ep_costs* provides the equivalent periodical costs for an investment, i.e., an output or input.

The terms in Eqs. 3.22 to 3.27 have been implemented in the programming code in order to calculate costs. The first constraint is conveyed in the term _investvar_bound_rule(block, i, o), where the variable investment is created.

```
def _investvar_bound_rule(block, i, o):
    return (m.flows[i, o].investment.minimum,
        m.flows[i, o].investment.maximum)
# Create variable bounded for flows with investement attribute.
# 'Var' represents 'Variable'.
self.invest = Var(self.FLOWS, within=NonNegativeReals,
        bounds=_investvar_bound_rule)
# The last sentence creates the variable 'invest'.
# The variable 'invest' must comply with the constraints set for this
# variable and stay within the set bounds. The constraint
# limits the variable to non-negative decimal point numbers (real numbers).
```

The next function _investflow_fixed_rule(block, i, o, t) creates rules for the investment flow in accordance with Eq. 3.22 (for the fixed investment flow):

```
def _investflow_fixed_rule(block, i, o, t):
    return (m.flow[i, o, t] == (self.invest[i, o] *
        m.flows[i, o].actual_value[t]))
self.fixed = Constraint(self.FIXED_FLOWS, m.TIMESTEPS,
        rule=_investflow_fixed_rule)
# The constraint 'fixed' belongs to the set 'FIXED_Flows'. For all
# flows for which the attribute 'fixed' is set to 'true'
# for the relevant timestep, the variable 'flow' is interpreted as equal to
# 'actual_value'. This changes the variable into a constant.
# Such a setup could be used for consumption or volatile energy feedback
# values depending on the model.
```

The following rules are defined for maximum and minimum investment flow in the programming code. The maximum investment flow is defined in Eq. 3.24:

```
def _max_investflow_rule(block, i, o, t):
    expr = (m.flow[i, o, t] <= (m.flows[i, o].max[t] *
        self.invest[i, o]))
    return expr
# The term 'expr' is a Python variable. This variable
# is issued in the next step.
self.max = Constraint(self.FLOWS, m.TIMESTEPS,
        rule=_max_investflow_rule)
```

This is the term for the minimum investment flow (cf. Eq. 3.23):

```
def _min_investflow_rule(block, i, o, t):
    expr = (m.flow[i, o, t] >= (m.flows[i, o].min[t] *
                self.invest[i, o]))
    return expr
self.min = Constraint(self.MIN_FLOWS, m.TIMESTEPS,
        rule=_min_investflow_rule)
```

For the sum parameters, rules for their maximum and minimum values are required. The maximum sum value is written in the following term in the source code (cf. Eq. 3.25):

```
def _summed_max_investflow_rule(block, i, o):
    """Rule definition for build action of max. sum flow constraint
    in investment case.
    """
    expr = (sum(m.flow[i, o, t] * m.timeincrement[t]
        for t in m.TIMESTEPS) <= m.flows[i, o].summed_max *
                self.invest[i, o])
    return expr
self.summed_max = Constraint(self.SUMMED_MAX_FLOWS,
            rule=_summed_max_investflow_rule)
```

This is the function _summed_min_investflow_rule(block, i, o) for the minimum sum value according to Eq. 3.26:

```
def _summed_min_investflow_rule(block, i, o):
    """Rule definition for build action of min. sum flow constraint
    in investment case.
    """
    expr = (sum(m.flow[i, o, t] * m.timeincrement[t]
        for t in m.TIMESTEPS) >= m.flows[i, o].summed_min *
                self.invest[i, o])
    return expr
self.summed_min = Constraint(self.SUMMED_MIN_FLOWS,
            rule=_summed_min_investflow_rule)
```

Next we present the objective function based on Eq. 3.27. This equation is only used for flows that contain the attribute class 'Investment':

```
def _objective_expression(self):
    if not hasattr(self, 'FLOWS'):
        return 0

    m = self.parent_block()
    investment_costs = 0

    for i, o in self.FLOWS:
        if m.flows[i, o].investment.ep_costs is not None:
            investment_costs += (self.invest[i, o] *
                        m.flows[i, o].investment.ep_costs)
        else:
            raise ValueError('Missing value for investment costs!')

    self.investment_costs = Expression(expr=investment_costs)
    return investment_costs
```

3.1.3   Class Bus(SimpleBlock)

A prerequisite for buses is that they must be balanced, i.e., the sum of all inputs must correspond exactly to the sum of all outputs. This is formulated in the following constraint (cf. Eq. 3.28) (oemof 2014c):

$$\sum_{i \in INPUTS(n)} flow(i, n, t) \cdot \tau = \sum_{o \in OUTPUTS(n)} flow(n, o, t) \cdot \tau \qquad (3.28)$$

This applies to:

$$\forall n \in BUSES \text{ and } \forall t \in TIMESTEPS$$

This equation is reflected in the function _busbalance_rule(block). This function calculates input and output of a bus for each timestep and ensures that these are always balanced:

```
def _busbalance_rule(block):
    for t in m.TIMESTEPS:
        for n in group:
            lhs = sum(m.flow[i, n, t] * m.timeincrement[t]
                    for i in I[n])
            rhs = sum(m.flow[n, o, t] * m.timeincrement[t]
                    for o in O[n])
            if expr = (lhs == rhs)
```

```
      #  no inflows no outflows yield: 0 == 0 which is True
      expr is not True:
         block.balance.add((n, t), expr)
self.balance = Constraint(group, noruleinit=True)
self.balance_build = BuildAction(rule=_busbalance_rule)
```

3.1.4   Class Transformer(SimpleBlock)

A transformer is represented by a node in OEMOF. By means of a conversion factor the flow of a transformer is adapted to a specified norm. Input and output of a transformer have a linear relation to each other (cf. Eq. 3.29) (oemof o.J.-c):

$$\frac{flow(i, n, t)}{conversion\_factor(n, i, t)} = \frac{flow(n, o, t)}{conversion\_factor(n, o, t)} \tag{3.29}$$

This applies to:

$\forall t \in TIMESTEPS, \ \forall n \in TRANSFORMER, \ \forall i \in INPUTS(n) \ \text{and} \ \forall o \in OUTPUTS(n)$

The order of the terms in brackets of Eq. 3.29 differs between the two sides of the equation. The left side describes the *flow* of the input (i) towards the component (n) for timestep t. The right side describes the *flow* from the component towards the output. The terms in brackets must therefore also be applied in that order.

The function *_input_output_relation(block)* calculates the output and input for each element in a group according to Eq. 3.29:

```
def _input_output_relation(block):
    for t in m.TIMESTEPS:
        for n in group:
            for o in out_flows[n]:
                for i in in_flows[n]:
# The instruction 'try' is used for identifying errors in Python.
                    try:
                        lhs = (m.flow[i, n, t] /
                            n.conversion_factors[i][t])
                        rhs = (m.flow[n, o, t] /
                            n.conversion_factors[o][t])
                    except ValueError:
                        raise ValueError(
                            'Error in constraint creation',
```

```
                    'source: {0}, target: {1}'.format(
                              n.label, o.label))
              block.relation.add((n, i, o, t), (lhs == rhs))
self.relation_build = BuildAction(rule=_input_output_relation)
```

3.1.5   Class NonConvexFlow(SimpleBlock)

The term *nonconvex* is a mathematical term and describes the nature of a function.
If a function is *convex*, this means that this function has an overall maximum or
minimum. *Nonconvex* functions have both an overall optimal level and several local
optimal levels (cf. Fig. 3.7). This is important to note when creating an optimization
function, as is the case with the OEMOF system of programs, as the solver will look
for an overall maximum or minimum.

Fig. 3.7 Example of a nonconvex function (based on Mistakidis 1998)

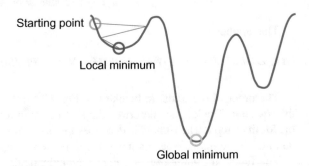

In OEMOF the nonconvex approach for functions is applied to the class *flow*.
The implementation of a nonconvex function in OEMOF is set up via several
constraints and via a dedicated objective function.

The first constraint defines the nonconvex flow (cf. Eq. 3.30)(oemof o.J.-c): For
this to be encoded, a new variable *status* is created that will take on binary values
(cf. Table 2.18). The parameter *nominal_value* is a numeric value representing a
nominal value.

$$flow(i, o, t) \geq \min(i, o, t) \cdot nominal\_value \cdot status(i, o, t) \qquad (3.30)$$

This applies to:

$$\forall t \in TIMESTEPS \text{ and } \forall (i, o) \in NONCONVEX\_FLOWS$$

An upper bound is also defined (cf. Eq. 3.31):

$$flow(i,o,t) \leq \max(i,o,t) \cdot nominal\_value \cdot status(i,o,t) \tag{3.31}$$

This applies to:

$$\forall t \in TIMESTEPS, \ \forall \ (i,o) \in NONCONVEX\_FLOWS$$

Starting up and powering down energy plants generates another type of costs than the operation of a power plant. For this reason, the startup of a component is formulated via the variable *startup*. A specific constraint is formulated for running the startup of a powerplant (cf. Eq. 3.32):

$$startup(i,o,t) \geq status(i,o,t) - status(i,o,t-1) \tag{3.32}$$

This applies to:

$$\forall t \in TIMESTEPS \ \text{and} \ \forall(i,o) \in STARTUP\_FLOWS$$

The variable *startup* then becomes part of the objective function.

Power plants, such as wind power plants, are powered down for various reasons, sometimes even when the weather situation would allow them to run. One reason might be low demand, i.e., an oversupply in the grid, another reason might be a low share value of the company. In some cases, if the wind speed is too high, this will also lead to a powering down of a wind turbine. Idle power plants generate costs. In order to calculate costs for idleness, the following constraint is formulated using the variable *shutdown* (cf. Eq. 3.33):

$$shutdown(i,o,t) \geq status(i,o,t-1) - status(i,o,t) \tag{3.33}$$

This applies to:

$$\forall t \in TIMESTEPS \ \text{and} \ \forall(i,o) \in SHUTDOWN\_FLOWS$$

The variable *shutdown* is also included in the objective function. This means that two terms for the objective function can be formulated for this class. But this only works if the modeler provides costs for starting up a power station (*startup_costs(i,o)*) and powering down a plant (*shutdown_costs(i,o)*). The first term for the objective function is presented in Eq. 3.34:

$$\sum_{i,o \in STARTUP\_FLOWS} \sum_{t} startup(i,o,t) \cdot startup\_costs(i,o) \tag{3.34}$$

The second term calculates the costs of powering down a plant (cf. Eq. 3.35):

$$\sum_{i,o \in SHUTDOWN\_FLOWS} \sum_t shutdown(i,o,t) \cdot shutdown\_costs(i,o) \qquad (3.35)$$

The programming code for these constraints and the objective function are spread across several methods and functions in OEMOF. The function _minimum_flow_rule(block, i, o, t) sets the rules for the flow minimum (cf. Eq. 3.30). This constraint is particularly relevant when the model is based on a mixed-integer linear optimization (MILP):

```
def _minimum_flow_rule(block, i, o, t):
    expr = (self.status[i, o, t] *
        m.flows[i, o].min[t] * m.flows[i, o].nominal_value <=
        m.flow[i, o, t])
    return expr
self.min = Constraint(self.MIN_FLOWS, m.TIMESTEPS,
        rule=_minimum_flow_rule)
```

The next constraint for a maximum flow limit based on Eq. 3.31 also applies to a MILP model:

```
def _maximum_flow_rule(block, i, o, t):
    expr = (self.status[i, o, t] *
        m.flows[i, o].max[t] * m.flows[i, o].nominal_value >=
        m.flow[i, o, t])
    return expr
self.max = Constraint(self.MIN_FLOWS, m.TIMESTEPS,
        rule=_maximum_flow_rule)
```

The function _startup_rule(block, i, o, t) defines the startup constraint according to Eq. 3.32:

```
def _startup_rule(block, i, o, t):
    if t > m.TIMESTEPS[1]:
        expr = (self.startup[i, o, t] >= self.status[i, o, t] -
            self.status[i, o, t-1])
    else:
        expr = (self.startup[i, o, t] >= self.status[i, o, t] -
            m.flows[i, o].nonconvex.initial_status)
    return expr
self.startup_constr = Constraint(self.STARTUPFLOWS, m.TIMESTEPS,
            rule=_startup_rule)
```

The constraint for a shutdown is reflected in the function _shutdown_rule(block, i, o, t) (cf. Eq. 3.33):

```
def _shutdown_rule(block, i, o, t):
    if t > m.TIMESTEPS[1]:
        expr = (self.shutdown[i, o, t] >= self.status[i, o, t-1] -
            self.status[i, o, t])
    else:
        expr = (self.shutdown[i, o, t] >=
            m.flows[i, o].nonconvex.initial_status -
            self.status[i, o, t])
    return expr
self.shutdown_constr = Constraint(self.SHUTDOWNFLOWS, m.TIMESTEPS,
                rule=_shutdown_rule)
```

Two new constraints have been added, one for the minimum running time (minimum uptime) (_min_uptime_rule(block, i, o, t)) and the minimum stopping time (minimum downtime) (_min_downtime_rule(block, i, o, t)):

```
def _min_uptime_rule(block, i, o, t):
    if (t >= m.flows[i, o].nonconvex.max_up_down and
        t <= m.TIMESTEPS[-1]-m.flows[i, o].nonconvex.max_up_down):
        expr = 0
        expr += ((self.status[i, o, t]-self.status[i, o, t-1]) *
            m.flows[i, o].nonconvex.minimum_uptime)
        expr += -sum(self.status[i, o, t+u] for u in range(0,
            m.flows[i, o].nonconvex.minimum_uptime))
        return expr <= 0
else:
    expr = 0
    expr += self.status[i, o, t]
    expr += -m.flows[i, o].nonconvex.initial_status
    return expr == 0
self.min_uptime_constr = Constraint(
    self.MINUPTIMEFLOWS, m.TIMESTEPS, rule=_min_uptime_rule)


def _min_downtime_rule(block, i, o, t):
    if (t >= m.flows[i, o].nonconvex.max_up_down and
        t <= m.TIMESTEPS[-1]-m.flows[i, o].nonconvex.max_up_down):
        expr = 0
        expr += ((self.status[i, o, t-1]-self.status[i, o, t]) *
            m.flows[i, o].nonconvex.minimum_downtime)
        expr += - m.flows[i, o].nonconvex.minimum_downtime
        expr += sum(self.status[i, o, t+d] for d in range(0,
```

```
                m.flows[i, o].nonconvex.minimum_downtime))
            return expr <= 0
        else:
            expr = 0
            expr += self.status[i, o, t]
            expr += -m.flows[i, o].nonconvex.initial_status
            return expr == 0
    self.min_downtime_constr = Constraint(
        self.MINDOWNTIMEFLOWS, m.TIMESTEPS, rule=_min_downtime_rule)
```

As part of the objective function the startup and the shutdown costs are added up. The following method *_objective_expression(self)* defines this process in Eqs. 3.34 and 3.35:

```
def _objective_expression(self):
    if not hasattr(self, 'NONCONVEX_FLOWS'):
        return 0

    m = self.parent_block()

    startcosts = 0
    shutdowncosts = 0

    if self.STARTUPFLOWS:
        for i, o in self.STARTUPFLOWS:
            if (m.flows[i, o].nonconvex.startup_costs[0] is not None):
                startcosts += sum(self.startup[i, o, t] *
                        m.flows[i, o].nonconvex.startup_costs[t]
                        for t in m.TIMESTEPS)
            self.startcosts = Expression(expr=startcosts)

    if self.SHUTDOWNFLOWS:
        for i, o in self.SHUTDOWNFLOWS:
            if (m.flows[i, o].nonconvex.shutdown_costs[0] is not None):
                shutdowncosts += sum(
                    self.shutdown[i, o, t] *
                    m.flows[i, o].nonconvex.shutdown_costs[t]
                    for t in m.TIMESTEPS)
            self.shutdowncosts = Expression(expr=shutdowncosts)

    return startcosts + shutdowncosts
```

3.2   Mathematical Equations in the Module components.py

This module describes equations and inequations for the following classes (GitHub 2017b):

- class *GenericStorage(network.Transformer)*
- class *GenericStorageBlock(SimpleBlock)*
- class *GenericInvestmentStorageBlock(SimpleBlock)*
- class *GenericCHP(network.Transformer)*
- class *GenericCHPBlock(SimpleBlock)*
- class *ExtractionTurbineCHP(solph_Transformer)*
- class *ExtractionTurbineCHPBlock(SimpleBlock)*

3.2.1   Class GenericStorage(network.Transformer)

This class only requires one method for the initialization of parameters. This method also carries out the verification of the parameters *investment, inputs.value()* and *outputs.value()* for flows and is implemented like this:

```
def __init__(self, *args, **kwargs):
        super().__init__(*args, **kwargs)


# The attribute 'kwargs' in the next line is a dictionary. 'kwargs'
# stands for 'keyword arguments'. This attribute takes on all
# keyword arguments, which are then passed on to the function.
# These are then stored in a dictionary. The method 'get' fetches
# the values of a parameter, which are stored in the dictionary.
# If the modeler has not provided values, default values are used.
# For the parameter 'capacity loss', the default value is '0'.
    self.nominal_capacity = kwargs.get('nominal_capacity')
    self.nominal_input_capacity_ratio = kwargs.get(
        'nominal_input_capacity_ratio', None)
    self.nominal_output_capacity_ratio = kwargs.get(
        'nominal_output_capacity_ratio', None)
    self.initial_capacity = kwargs.get('initial_capacity')
    self.capacity_loss = solph_sequence(kwargs.get('capacity_loss', 0))
    self.inflow_conversion_factor = solph_sequence(
      kwargs.get(
          'inflow_conversion_factor', 1))
    self.outflow_conversion_factor = solph_sequence(
      kwargs.get(
          'outflow_conversion_factor', 1))
```

```python
self.capacity_max = solph_sequence(kwargs.get('capacity_max', 1))
self.capacity_min = solph_sequence(kwargs.get('capacity_min', 0))
self.investment = kwargs.get('investment')

# General error messages
e_no_nv = ("If an investment object is defined the invest variable "
      "replaces the {0}.\n Therefore the {0} should be "
      "'None'.\n'')
e_duplicate = (
  "Duplicate definition.\nThe 'nominal_{0}_capacity_ratio'"
  "will set the nominal_value for the flow.\nTherefore "
  "either the 'nominal_{0}_capacity_ratio' or the "
  "'nominal_value' has to be 'None'.")
# Check investment
if self.investment and self.nominal_capacity is not None:
    raise AttributeError(e_no_nv.format('nominal_capacity'))

# Check input flows
for flow in self.inputs.values():
    if self.investment and flow.nominal_value is not None:
      raise AttributeError(e_no_nv.format('nominal_value'))
    if (flow.nominal_value is not None and
        self.nominal_input_capacity_ratio is not None):
      raise AttributeError(e_duplicate)
    if (not self.investment and
        self.nominal_input_capacity_ratio is not None):
     flow.nominal_value = (self.nominal_input_capacity_ratio *
              self.nominal_capacity)
    if self.investment:
      if not isinstance(flow.investment, Investment):
        flow.investment = Investment()

# Check output flows
for flow in self.outputs.values():
    if self.investment and flow.nominal_value is not None:
     raise AttributeError(e_no_nv.format('nominal_value'))
    if (flow.nominal_value is not None and
        self.nominal_output_capacity_ratio is not None):
     raise AttributeError(e_duplicate)
    if (not self.investment and
        self.nominal_output_capacity_ratio is not None):
     flow.nominal_value = (self.nominal_output_capacity_ratio *
              self.nominal_capacity)
```

```
if self.investment:
    if not isinstance(flow.investment, Investment):
        flow.investment = Investment()
```

The following section lists values for the required attributes for this class:

```
from oemof import solph
    my_bus = solph.Bus('my_bus')
    my_storage = solph.components.GenericStorage(
        label='storage',
        nominal_capacity=1000,
        inputs={my_bus: solph.Flow(variable_costs=10)},
        outputs={my_bus: solph.Flow()},
        capacity_loss=0.01,
        initial_capacity=0,
        capacity_max = 0.9,
        nominal_input_capacity_ratio=1/6,
        nominal_output_capacity_ratio=1/6,
        inflow_conversion_factor=0.9,
        outflow_conversion_factor=0.93)
    my_investment_storage = solph.components.GenericStorage(
        label='storage',
        investment=solph.Investment(ep_costs=50),
        inputs={my_bus: solph.Flow()},
        outputs={my_bus: solph.Flow()},
        capacity_loss=0.02,
        initial_capacity=None,
        nominal_input_capacity_ratio=1/6,
        nominal_output_capacity_ratio=1/6,
        inflow_conversion_factor=1,
        outflow_conversion_factor=0.8)
```

3.2.2   Class GenericStorageBlock(SimpleBlock)

The class energy storage is also defined as balanced. This is ensured by the following constraint (cf. Eq. 3.36) (oemof o.J.-c):

$$
\begin{aligned}
capacity(n, t) = {} & capacity(n, previous(t)) \cdot (1 - capacit\_loss_n(t)) \\
& - \frac{flow(n, o, t)}{\eta(n, o, t)} \cdot \tau + flow(i, n, t) \cdot \eta(i, n, t) \cdot \tau
\end{aligned}
\tag{3.36}
$$

The term *capacity* is a variable which indicates the level of an energy storage *n* per timestep *t*. At the beginning and at the end of the modeling, this value is set to *initial_capacity* unless this value is *None*. The parameter *initial_capacity* is defined as numeric. This parameter sets an initial capacity for the storage unit *n* for the first and for the last timestep. There is also the parameter *capacity_loss* for storage *n*, which describes the losses incurred between timesteps. The term η either represents the *inflow_conversion_factor* or the *outflow_conversion_factor*. Both are numeric values.

Further constraints are formulated. The first constraint sets a minimum and a maximum storage capacity by means of the function _storage_capac-ity_bound_rule(block, n, t):

```
def _storage_capacity_bound_rule(block, n, t):
# The letter 'n' represents storage unit n and t in the timestep.
# This is used to calculate storage capacity for storage unit n
# in timestep t.
    bounds = (n.nominal_capacity * n.capacity_min[t],
            n.nominal_capacity * n.capacity_max[t])
    return bounds
self.capacity = Var(self.STORAGES, m.TIMESTEPS,
            bounds=_storage_capacity_bound_rule)


# set the initial capacity of the storage
for n in group:
    if n.initial_capacity is not None:
        self.capacity[n, m.TIMESTEPS[-1]] = (n.initial_capacity *
                            n.nominal_capacity)
        self.capacity[n, m.TIMESTEPS[-1]].fix()
```

The function _storage_balance_rule(block, n, t) defines the rules for calculating the energy balance in the storage unit. This is again calculated for each storage unit *n* in timestep *t*:

```
# Storage balance constraint
def _storage_balance_rule(block, n, t):
    expr = 0
    expr += block.capacity[n, t]
    expr += - block.capacity[n, m.previous_timesteps[t]] * (
        1 - n.capacity_loss[t])
    expr += (- m.flow[i[n], n, t] *
        n.inflow_conversion_factor[t]) * m.timeincrement[t]
```

```
    expr += (m.flow[n, o[n], t] /
        n.outflow_conversion_factor[t]) * m.timeincrement[t]
    return expr == 0
self.balance = Constraint(self.STORAGES, m.TIMESTEPS,
            rule=_storage_balance_rule)
```

3.2.3   Class GenericInvestmentStorageBlock(SimpleBlock)

This class represents the calculation of investment costs (oemof o.J.-c). First, the energy balance of a storage unit is calculated across a time period (cf. Eq. 3.37):

$$capacity(n, t) = capacity(n, t\_previous(t)) \cdot (1 - capacity\_loss(n)) - (flow(n, target(n), t))/$$
$$(outflow\_conversion\_factor(n) \cdot \tau) + flow(souce(n), n, t) \cdot inflow\_conversion\_factor(n) \cdot \tau$$

$$(3.37)$$

This applies to:

$$\forall n \in INVESTSTORAGES \;\; and \;\; \forall T \in TIMESTEPS$$

The rules for the energy balance of a storage unit are defined by the function _storage_balance_rule(block, n, t):

```
def _storage_balance_rule(block, n, t):
    expr = 0
    expr += block.capacity[n, t]
    expr += - block.capacity[n, m.previous_timesteps[t]] * (
        1 - n.capacity_loss[t])
    expr += (- m.flow[i[n], n, t] *
        n.inflow_conversion_factor[t]) * m.timeincrement[t]
    expr += (m.flow[n, o[n], t] /
        n.outflow_conversion_factor[t]) * m.timeincrement[t]
    return expr == 0
self.balance = Constraint(self.INVESTSTORAGES, m.TIMESTEPS,
            rule=_storage_balance_rule)
```

The capacity of the last timestep is calculated by means of the constraint in Eq. 3.38:

$$capacity(n, t_{last}) = invest(n) \cdot initial\_capacity(n) \qquad (3.38)$$

This applies to:

$$\forall n \in INITIAL\_CAPACITY \text{ and } \forall t \in TIMESTEPS$$

In Eq. 3.38 the parameter *initial_capacity(n)* represents the capacity of a storage unit for the last timestep of the optimization function. This parameter is also numeric. Furthermore, this equation contains two variables. The first variable is $capacity(n, t_{last})$. This describes the storage level at the end of the time period. The second variable *invest(n)* provides the nominal capacity of the storage (invested capacity).

The next function *_initial_capacity_invest_rule(block, n)* defines the rules for connecting the capacity at the beginning and at the end of the time period:

```
def _initial_capacity_invest_rule(block, n):
    expr = (self.capacity[n, m.TIMESTEPS[-1]] == (n.initial_capacity *
        self.invest[n]))
    return expr
self.initial_capacity = Constraint(
    self.INITIAL_CAPACITY, rule=_initial_capacity_invest_rule)
```

Next the variables for the investment are first connected to the input flow (cf. Eq. 3.39) and then to the output flow (cf. Eq. 3.40):

$$InvestmentFlow.invest(source(n), n) = invest(n) \cdot nominal\_input\_capacity\_ratio(n)$$

$$(3.39)$$

This applies to:

$$\forall n \in INVESTSTORAGES$$

The variable *investmentflow.invest* provides the value of the investment variable. After optimization this will be equivalent to the nominal value of the flow. The parameter *nominal_input_capacity_ratio* defines the ratio between the nominal inflow of the storage and its capacity. The set *INVESTSTORAGES* contains all investment objects.

Let us first look at the input flow described by Eq. 3.39. This is defined in the function: *_storage_capacity_inflow_invest_rule(block, n)*:

```
def _storage_capacity_inflow_invest_rule(block, n):
    """Rule definition of constraint connecting the inflow
    'InvestmentFlow.invest' of storage with invested capacity 'invest'
    by nominal_capacity__inflow_ratio
    """
```

```
    expr = (m.InvestmentFlow.invest[i[n], n] ==
        self.invest[n] * n.nominal_input_capacity_ratio)
    return expr
self.storage_capacity_inflow = Constraint(
    self.INVESTSTORAGES, rule=_storage_capacity_inflow_invest_rule)
```

The relation to the output is defined like this (cf. Eq. 3.40):

$$InvestmentFlow.invest(n, target(n)) = invest(n) \cdot nominal\_output\_capacity\_ratio(n)$$
$$(3.40)$$

This applies to:

$$\forall n \in INVESTSTORAGES$$

Find here below the function _storage_capacity_outflow_invest_rule(block, n) according to Eq. 3.40 for the output flow:

```
def _storage_capacity_outflow_invest_rule(block, n):
    expr = (m.InvestmentFlow.invest[n, o[n]] ==
        self.invest[n] * n.nominal_output_capacity_ratio)
    return expr
self.storage_capacity_outflow = Constraint(
    self.INVESTSTORAGES, rule=_storage_capacity_outflow_invest_rule)
```

What is still required are constraints for the maximum and the minimum capacity. For the capacity not to exceed the maximum capacity, the following constraint is formulated in Eq. 3.41:

$$capacity(n, t) \leq invest(n) \cdot capacity\_max(n, t)$$
$$(3.41)$$

This applies to:

$$\forall n \in MAX\_INVESTSTORAGES \text{ and } \forall t \in TIMESTEPS$$

The variable *capacity* provides the storage level. For the parameter *capacity_-max* a numeric value must be set for the nominal maximum capacity of the storage, described as a fraction of the nominal capacity. This value must be between 0 and 1. The default value is 0.

The function _max_capacity_invest_rule(block, n, t) sets an upper limit for storage capacity. The source code for defining the upper limit looks like this:

```
def _max_capacity_invest_rule(block, n, t):
    expr = (self.capacity[n, t] <= (n.capacity_max[t] *
        self.invest[n]))
    return expr
self.max_capacity = Constraint(
    self.INVESTSTORAGES, m.TIMESTEPS, rule=_max_capacity_invest_rule)
```

In order to ensure that the actual capacity is above the minimum level at any given point in time, the following constraint is formulated in Eq. 3.42:

$$capacity(n,t) \geq invest(n) \cdot capacity\_min(n,t) \tag{3.42}$$

This applies to:

$$\forall n \in MIN\_INVESTSTORAGES \text{ and } \forall t \in TIMESTEPS$$

The parameters, i.e., variables for this have already been introduced. For the parameter *capacity_min* a minimum value must be entered.

The minimum bound for capacity is added to the model via the function *_min_capacity_invest_rule(block, n, t)* in Eq. 3.42. It has a similar structure to the function for the upper limit:

```
def _min_capacity_invest_rule(block, n, t):
    expr = (self.capacity[n, t] >= (n.capacity_min[t] *
                        self.invest[n]))
    return expr
# Set the lower bound of the storage capacity if the attribute exists
self.min_capacity = Constraint(
    self.MIN_INVESTSTORAGES, m.TIMESTEPS,
    rule=_min_capacity_invest_rule)
```

The contribution of this class to the objective function is described in Eq. 3.43. This is used to calculate the equivalent periodical costs (investment costs):

$$sum_n invest(n) \cdot ep\_costs(n) \tag{3.43}$$

For calculating this, first the sum of all storage units is added up. The term *ep_costs* contains a parameter which provides the equivalent periodical costs for the investment as a floating decimal point number. If the optimization model is run across a whole year these costs correspond to the equivalent annual costs.

The objective function itself is contained in the method *_objective_expression (self)*:

```
def _objective_expression(self):
    if not hasattr(self, 'INVESTSTORAGES'):
        return 0

    investment_costs = 0

    for n in self.INVESTSTORAGES:
        if n.investment.ep_costs is not None:
            investment_costs += self.invest[n] * n.investment.ep_costs
        else:
            raise ValueError('Missing value for investment costs!')

    self.investment_costs = Expression(expr=investment_costs)

    return investment_costs
```

3.2.4   Class GenericCHP(network.Transformer)

The modeling of the GenericCHP(network.Transformer) component is based on Mollenhauer (2016). Further information about a MILP model for a CHP plant can be found in Salgado (2008).

CHP plants have become particularly important after the energy transition in Germany. As the actual processes in these plants are very complex, simplified models based on mixed-integer linear programs (MILP) have been developed. These are used in energy market models or price-based unit commitment problems. These can provide feasibility studies or optimal operation strategies for different plants (Mollenhauer 2016).

The model described in Mollenhauer (2016) can be used to describe steady state characteristics of CHP plants. The problems solved by these models tend to be based on optimal operation planning for power plants. The function used to describe this must contain dynamic constraints, such as startup costs, maximum ramp rates, and minimum operation and down time intervals.

The most important equations for this are described in Sect. 3.2.5.

In Mollenhauer (2016) two alpha coefficients α_1 and α_2 are defined in order to describe the efficiency η_{el} of a CHP plant. The electrical efficiency for CHP plants is normally calculated by means of Eq. 3.44:

$$\eta_{el} = \frac{P}{\dot{H}_F} \qquad (3.44)$$

where

P power output

$\dot{H}_F$ energy level of the fuel used

η_{el} electrical efficiency

By means of the two alpha coefficients the electrical efficiency can be calculated based on boiler load. In order to arrive at coefficents α_1 and α_2, the following two equations Eq. 3.45 and 3.46 must be solved:

$$\eta_{el,min,woDH} = \frac{P_{min,woDH}}{\alpha_1 + \alpha_2 \cdot P_{min,woDH}} \tag{3.45}$$

$$\eta_{el,max,woDH} = \frac{P_{max,woDH}}{\alpha_1 + \alpha_2 \cdot P_{max,woDH}} \tag{3.46}$$

where

$\eta_{el,min,woDH}$ minimum electrical efficiency without heat extraction
$\eta_{el,max,woDH}$ maximum electrical efficiency without heat extraction
$P_{min,woDH}$ minimum electrical power without heat extraction
$P_{max,woDH}$ maximum electrical power without heat extraction
α_1, α_2 coefficients

The two alpha coefficients α_1 and α_2 are required for further calculations in Sect. 3.2.5.

By means of the method (_calculate_alphas(self)) the two alpha coefficients are calculated in OEMOF for this class:

```
def _calculate_alphas(self):
    alphas = [[], []]


# The term 'eb' stands for 'Elastic Beanstalk' [AWS 2018]. This is a
# specific Python variable. Elastic Beanstalk provides the resources
# required for executing an application.
    eb = list(self.electrical_output.keys())[0]

    attrs = [self.electrical_output[eb].P_min_woDH,
        self.electrical_output[eb].Eta_el_min_woDH,
        self.electrical_output[eb].P_max_woDH,
        self.electrical_output[eb].Eta_el_max_woDH]

    length = [len(a) for a in attrs if not isinstance(a, (int, float))]
    max_length = max(length)

    if all(len(a) == max_length for a in attrs):
        if max_length == 0:
```

```
      max_length += 1 # increment dimension for scalars from 0 to 1
    for i in range(0, max_length):
      A = np.array([[1, self.electrical_output[eb].P_min_woDH[i]],
                [1, self.electrical_output[eb].P_max_woDH[i]]])
      b = np.array([self.electrical_output[eb].P_min_woDH[i] /
                self.electrical_output[eb].Eta_el_min_woDH[i],
                self.electrical_output[eb].P_max_woDH[i] /
                self.electrical_output[eb].Eta_el_max_woDH[i]])
      x = np.linalg.solve(A, b)
      alphas[0].append(x[0])
      alphas[1].append(x[1])
    else:
      error_message = ('Attributes to calculate alphas ' +
                'must be of same dimension.')
      raise ValueError(error_message)

    self._alphas = alphas
```

We will here look at some attributes in more detail:

```
>>> from oemof import solph
  >>> bel = solph.Bus(label='electricityBus')
  >>> bth = solph.Bus(label='heatBus')
  >>> bgas = solph.Bus(label='commodityBus')
  >>> ccet = solph.components.GenericCHP(
  ...   label='combined_cycle_extraction_turbine',
  ...   fuel_input={bgas: solph.Flow(
  ...     H_L_FG_share_max=[0.183])},
  ... electrical_output={bel: solph.Flow(
  ...   P_max_woDH=[155.946],
  ...   P_min_woDH=[68.787],
  ...   Eta_el_max_woDH=[0.525],
  ...   Eta_el_min_woDH=[0.444])},
  ... heat_output={bth: solph.Flow(
  ...   Q_CW_min=[10.552])},
  ... Beta=[0.122], back_pressure=False)
  >>> type(ccet)
  <class 'oemof.solph.components.GenericCHP'>
```

3.2.5   Class *GenericCHPBlock(SimpleBlock)*

This section looks at the equations and inequations for modeling a CHP plant. OEMOF provides these only in a separate file (see GitHub 2017b). The class *GenericCHPBlock(SimpleBlock)* is used to group all CHP plants together, so that these can be modeled and calculated together.

For modeling the operation of a CHP plant, certain constraints are needed for calculating the fuel required and the heat and the electricity that are produced. A description of the mathematical and thermodynamic basis of these can be found in Mollenhauer (2016).

Fuel is put into the plant. For calculating the energy content of the fuel input (H_F) the lower heat value (H_u) and its mass flux ($\dot{m}_F$) are used (cf. Eq. 3.47):

$$\dot{H}_F = H_u \cdot \dot{m}_F \tag{3.47}$$

In the source code this constraint is formulated in the following function: *_H_flow_rule(block, n, t)*. For the calculation the values for energy content are read from a pre-prepared list.

```
def _H_flow_rule(block, n, t):
    expr = 0
    expr += self.H_F[n, t]
    expr += - m.flow[list(n.fuel_input.keys())[0], n, t]
    return expr == 0
self.H_flow = Constraint(self.GENERICCHPS, m.TIMESTEPS,
            rule=_H_flow_rule)
```

After the energy has been transformed, heat is the output of the plant. The following code section containing the function *_Q_flow_rule(block, n, t)* relates the heat produced to the output of the plant via its flow:

```
def _Q_flow_rule(block, n, t):

    expr = 0
    expr += self.Q[n, t]
    expr += - m.flow[n, list(n.heat_output.keys())[0], t]
    return expr == 0
self.Q_flow = Constraint(self.GENERICCHPS, m.TIMESTEPS,
            rule=_Q_flow_rule)
```

Another output of the plant is the electricity produced by the plant, which must also be related to the plant via a flow. This is carried out by the function *_P_flow_rule(block, n, t)*:

```
def _P_flow_rule(block, n, t):

    expr = 0
    expr += self.P[n, t]
    expr += - m.flow[n, list(n.electrical_output.keys())][0], t]
    return expr == 0
self.P_flow = Constraint(self.GENERICCHPS, m.TIMESTEPS,
            rule=_P_flow_rule)
```

Equation 3.48 can be used to describe the load-dependent efficiency in MILP:

$$\dot{H}_F(t) = \alpha_1 \cdot Y(t) + \alpha_2 \cdot P_{\text{woDH}} \tag{3.48}$$

This applies to:

$$\forall t$$

where

$Y(t)$ binary status variable

The formula for calculating the two coefficients α_1 and α_2 were already introduced in equations Eqs. 3.44–3.46:

In the source code Eq. 3.48 is formulated in the following way:

```
def _H_F_1_rule(block, n, t):

    expr = 0
    expr += - self.H_F[n, t]
    expr += n.alphas[0][t] * self.Y[n, t]
    expr += n.alphas[1][t] * self.P_woDH[n, t]
    return expr == 0
self.H_F_1 = Constraint(self.GENERICCHPS, m.TIMESTEPS,
            rule=_H_F_1_rule)
```

There are interdependencies between the fuel input, the electricity output and the heat output. In order to describe these mathematically, Eq. 3.48 is extended to include the heat generation by means of the power loss coefficient β (cf. Eq. 3.49):

$$\dot{H}_F(t) = \alpha_1 \cdot Y(t) \\ + \alpha_2 \cdot \left(P(t) + \beta(T_{\text{FF}}(t), T_{\text{RF}}(t), T_0(t)) \cdot \dot{Q}(t) \right) \tag{3.49}$$

This applies to:

$$\forall t$$

where

$\dot{Q}$ heat output

The coefficient β is set for each timestep t with a feed temperature (T_{FF}) and return flow (T_{RF}) temperature in order to reach a predefined room temperature. Temperature T_0 is the reference temperature. The power loss coefficient β is calculated based on the thermodynamics equation Eq. 3.50

$$\beta = \left(1 - \frac{T_0}{T_M}\right) \tag{3.50}$$

Temperature T_0 is the reference temperature for the exergy calculation. This should correspond to the cooling water inlet temperature T_M. Temperature T_M represents the median logarithmic temperature, which can be calculated according to Windisch (2014) and Mollenhauer (2016). This equation is represented in the source code as the function _H_F_2_rule(block, n, t):

```
def _H_F_2_rule(block, n, t):

    expr = 0
    expr += - self.H_F[n, t]
    expr += n.alphas[0][t] * self.Y[n, t]
    expr += n.alphas[1][t] * (self.P[n, t] + n.Beta[t] * self.Q[n, t])
    return expr == 0
self.H_F_2 = Constraint(self.GENERICCHPS, m.TIMESTEPS,
            rule=_H_F_2_rule)
```

There is an operational gap between the minimum boiler load and the powering down of a plant. There is also an upper operation limit for each plant. For this reason both a maximum and a minimum operation level must be set: The maximum operation level is defined by (cf. Eq. 3.51):

$$\dot{H}_F(t) \leq Y(t) \cdot \frac{P_{max,woDH}}{\eta_{el,max,woDH}} \tag{3.51}$$

This applies to:

$$\forall t$$

In the program code this is described by the function _H_F_3_rule(block, n, t):

```
def _H_F_3_rule(block, n, t):

    expr = 0
    expr += self.H_F[n, t]
    expr += - self.Y[n, t] * \
            (list(n.electrical_output.values())[0].P_max_woDH[t] /
            list(n.electrical_output.values())[0].Eta_el_max_woDH[t])
    return expr <= 0
self.H_F_3 = Constraint(self.GENERICCHPS, m.TIMESTEPS,
                rule=_H_F_3_rule)
```

The lower operation limit is described by this equation (cf. Eq. 3.52):

$$\dot{H}_F(t) \geq Y(t) \cdot \frac{P_{min,woDH}}{\eta_{el,min,woDH}} \tag{3.52}$$

This applies to:

$$\forall t$$

This relationship is described in the function _H_F_4_rule(block, n, t):

```
def _H_F_4_rule(block, n, t):

    expr = 0
    expr += self.H_F[n, t]
    expr += - self.Y[n, t] * \
            (list(n.electrical_output.values())[0].P_min_woDH[t] /
            list(n.electrical_output.values())[0].Eta_el_min_woDH[t])
    return expr >= 0
self.H_F_4 = Constraint(self.GENERICCHPS, m.TIMESTEPS,
                rule=_H_F_4_rule)
```

In a CHP plant there are also losses through the flue gas, which must also be described. A maximum level of this loss must be defined. This depends on the fuel used and is calculated by means of the function _H_L_FG_max_rule(block, n, t) in OEMOF.

```
def _H_L_FG_max_rule(block, n, t):
    expr = 0
    expr += - self.H_L_FG_max[n, t]
    expr += self.H_F[n, t] * \
            list(n.fuel_input.values())[0].H_L_FG_share_max[t]
    return expr == 0
self.H_L_FG_max_def = Constraint(self.GENERICCHPS, m.TIMESTEPS,
```

```
rule=_H_L_FG_max_rule)
```

The heat produced in a CHP plant also depends on the fuel used and the amount of electricity it produces. For the modeling it is important to set a maximum level for heat output (cf. Eq. 3.53):

$$P(t) \leq \dot{H}_F(t) - \dot{Q}(t) - \dot{H}_{L,FG}(t) - \dot{Q}_{CW,min} \cdot Y(t) \tag{3.53}$$

This applies to:

$$\forall t$$

where

$\dot{H}_{L,FG}(t)$ flue gas loss

$\dot{Q}_{CW,min}$ minimum heat transport in the capacitor

In the source code this is formulated by means of the function _Q_max_res_rule (block, n, t):

```
def _Q_max_res_rule(block, n, t):

    expr = 0
    expr += self.P[n, t] + self.Q[n, t] + self.H_L_FG_max[n, t]
    expr += list(n.heat_output.values())[0].Q_CW_min[t] * self.Y[n, t]
    expr += - self.H_F[n, t]
    # back-pressure characteristics or one-segment model
    if n.back_pressure is True:
        return expr == 0
    else:
        return expr <= 0
self.Q_max_res = Constraint(self.GENERICCHPS, m.TIMESTEPS,
                rule=_Q_max_res_rule)
```

The value $\dot{Q}_{CW,min}$ is the minimum heat transport in the capacitor, which is dependent on the flow conditions between the turbines. The minimum heat transport can be set at 10% of the nominal capacity of a turbine without steam extraction (Mollenhauer 2016). The value $\dot{Q}_{CW,min}$ can be calculated by Eq. 3.54:

$$\dot{Q}_{CW,min} = 10\% \cdot \left(\dot{H}_{F,max} - \dot{H}_{L,FG} - P_{max,woDH} \right) \tag{3.54}$$

For the flue gas loss $\dot{H}_{L,FG}(t)$ the function _H_L_FG_min_rule(block, n, t) describes the minimum level:

```
def _H_L_FG_min_rule(block, n, t):
    if getattr(list(n.fuel_input.values())[0],
        'H_L_FG_share_min', None):
        expr = 0
        expr += - self.H_L_FG_min[n, t]
        expr += self.H_F[n, t] * \
            list(n.fuel_input.values())[0].H_L_FG_share_min[t]
        return expr == 0
    else:
        return Constraint.Skip
self.H_L_FG_min_def = Constraint(self.GENERICCHPS, m.TIMESTEPS,
                    rule=_H_L_FG_min_rule)
```

The minimum heat production is also described dependent on the fuel used and the electricity produced by the function _Q_min_res_rule(block, n, t):

```
def _Q_min_res_rule(block, n, t):
    if getattr(list(n.fuel_input.values())[0],
        'H_L_FG_share_min', None):
        expr = 0
        expr += self.P[n, t] + self.Q[n, t] + self.H_L_FG_min[n, t]
        expr += list(n.heat_output.values())[0].Q_CW_min[t] \
            * self.Y[n, t]
        expr += - self.H_F[n, t]
        return expr >= 0
    else:
        return Constraint.Skip
self.Q_min_res = Constraint(self.GENERICCHPS, m.TIMESTEPS,
                rule=_Q_min_res_rule)
```

3.2.6   Class ExtractionTurbineCHP(solph_Transformer)

The thermodynamic calculation bases and constraints for the class *ExtractionTurbineCHP(solph_Transformer)* correspond to the equations and inequations introduced in Sects. 3.2.4 and 3.2.5. Further equations can be found in Sect. 3.2.7.

The following source code excerpt gives a general description of how objects are created in this class (GitHub 2017b):

```
>>> from oemof import solph
    >>> bel = solph.Bus(label='electricityBus')
    >>> bth = solph.Bus(label='heatBus')
```

```
>>> bgas = solph.Bus(label='commodityBus')
>>> et_chp = solph.components.ExtractionTurbineCHP(
...     label='variable_chp_gas',
...     inputs={bgas: solph.Flow(nominal_value=10e10)},
...     outputs={bel: solph.Flow(), bth: solph.Flow()},
...     conversion_factors={bel: 0.3, bth: 0.5},
...     conversion_factor_full_condensation={bel: 0.5})
```

3.2.7   Class ExtractionTurbineCHPBlock(SimpleBlock)

Two constraints are defined in this class to describe the variable interdependency between different steam flows; steam input, steam output and steam extraction (tapped output) flows of a cogeneration plant. Electricity is produced by the turbine and heat is produced via steam extraction via the main steam output.

First the relation between the input flow to the two output flows (main output flow and tapped output flow) is described by means of the condensation efficiency (*efficiency_condensing*(n, t)) according to Eq. 3.55 (oemof o.J.-c):

$$
flow(input, n, t)
= \frac{(flow(n, main\_output, t) + flow(n, tapped\_output, t) \cdot main\_flow\_loss\_index(n, t)}{efficiency\_condensing(n, t)}
$$

$$(3.55)$$

This applies to:

$$\forall t \in TIMESTEPS \text{ and } \forall n \in VARIABLE\_FRACTION\_TRANSFORMERS$$

The efficiency rate *efficiency_condensing*(n, t) is a conversion factor for full condensation, which is formulated in the source code as *conversion_factor_full_condensation*.

Equation 3.56 defines a constraint for the relation between the two output types by means of the conversion factor if some steam extraction takes place (*conversion_factor*$(n, tapped\_output, t)$):

$$
flow(n, main\_output, t) = \frac{flow(n, tapped\_output, t) \cdot conversion\_factor(n, main\_output, t)}{conversion\_factor(n, tapped\_output, t)}
$$

$$(3.56)$$

This applies to:

$$\forall t \in TIMESTEPS \ and \ \forall n \in VARIABLE\_FRAKTION\_TRANSFORMERS$$

Conversion factors must be defined for both full CHP mode and condensing mode.

The two constraints of Eqs. 3.55 and 3.56 have been integrated into the following function. The constraints are created by means of _create (self,group=None). First the two methods and then the two functions are presented separately in the source code below:

```
def _create(self, group=None):

    Parameters
    --------------
    group : list
        List of :class:'oemof.solph.ExtractionTurbineCHP' (trsf)
        objects for
        which the linear relation of inputs and outputs is created
        e.g. group = [trsf1, trsf2, trsf3, ...]. Note that the relation
        is created for all existing relations of the inputs and all outputs
        of the transformer. The components inside the list need to hold
        all needed attributes.
    " " "
    if group is None:
        return None

    m = self.parent_block()

    for n in group:
        n.inflow = list(n.inputs)[0]
        n.label_main_flow = str(
        [k for k, v in n.conversion_factor_full_condensation.items()]
        [0])
        n.main_output = [o for o in n.outputs
                if n.label_main_flow == o.label][0]
        n.tapped_output = [o for o in n.outputs
                    if n.label_main_flow != o.label][0]
        n.conversion_factor_full_condensation_sq = (
          n.conversion_factor_full_condensation[
            m.es.groups[n.main_output.label]])
        n.flow_relation_index = [
          n.conversion_factors[m.es.groups[n.main_output.label]][t] /
          n.conversion_factors[m.es.groups[n.tapped_output.label]][t]
          for t in m.TIMESTEPS]
```

```
n.main_flow_loss_index = [
    (n.conversion_factor_full_condensation_sq[t] -
    n.conversion_factors[m.es.groups[n.main_output.label]][t]) /
    n.conversion_factors[m.es.groups[n.tapped_output.label]][t]
    for t in m.TIMESTEPS]
```

The first function *_input_output_relation_rule(block)* defines the constraints of the relations between input and output (main flow and tapped output flow) according to Eq. 3.55:

```
def _input_output_relation_rule(block):

    for t in m.TIMESTEPS:
        for g in group:
            lhs = m.flow[g.inflow, g, t]
            rhs = (
                (m.flow[g, g.main_output, t] +
                m.flow[g, g.tapped_output, t] *
                g.main_flow_loss_index[t]) /
                g.conversion_factor_full_condensation_sq[t]
                )
            block.input_output_relation.add((n, t), (lhs == rhs))
self.input_output_relation = Constraint(group, noruleinit=True)
self.input_output_relation_build = BuildAction(
    rule=_input_output_relation_rule)
```

The second function *_out_flow_relation_rule(block)* describes the relations between the two different output flows (main flow and tapped output flow) in CHP mode according to Eq. 3.56:

```
def _out_flow_relation_rule(block):
    for t in m.TIMESTEPS:
        for g in group:
            lhs = m.flow[g, g.main_output, t]
            rhs = (m.flow[g, g.tapped_output, t] *
                g.flow_relation_index[t])
            block.out_flow_relation.add((g, t), (lhs >= rhs))
self.out_flow_relation = Constraint(group, noruleinit=True)
self.out_flow_relation_build = BuildAction(
    rule=_out_flow_relation_rule)
```

Another method *component_grouping(node)* sets up the grouping for this class:

```
def component_grouping(node):
    if isinstance(node, GenericStorage) and isinstance(node.investment,
                            Investment):
        return GenericInvestmentStorageBlock
    if isinstance(node, GenericStorage) and not
            isinstance(node.investment, Investment):
        return GenericStorageBlock
    if isinstance(node, GenericCHP):
        return GenericCHPBlock
    if isinstance(node, ExtractionTurbineCHP):
        return ExtractionTurbineCHPBlock
```

3.3   Mathematical Equation System in the Module constraints.py

This module contains constraints for emissions and investments.

This module provides the option to set an upper limit for investment (cf. Eq. 3.57):

$$\sum_{investment\_costs} \leq limit \tag{3.57}$$

The parameter *limit* provides an investment limit in the form of a floating decimal point number. In the source code this is executed by the method *investment_limit(model, limit=None)*:

```
def investment_limit(model, limit=None):

    def investment_rule(m):
        expr = 0

        if hasattr(m, "InvestmentFlow"):
            expr += m.InvestmentFlow.investment_costs

        if hasattr(m, "GenericInvestmentStorageBlock"):
            expr += m.GenericInvestmentStorageBlock.investment_costs
        return expr <= limit

model.investment_limit = po.Constraint(rule=investment_rule)

    return model
```

The method *investment_limit(model, limit=None)* provides a global investment limit. By means of the function *investment_rule(m)* it is possible to make this constraint apply to all components, so the limit does not have to be defined separately for each component. For more information on this see [GitHub 2017d].

The following method is used to set an upper limit for emissions *emission_limit (om, flows=None, limit=None)*:

```python
def emission_limit(om, flows=None, limit=None):
    """Set a global limit for emissions. The emission attribute must be
    added to every flow you want to consider.
    Parameters
    --------------
    om : oemof.solph.Model
        Model to which constraints are added.
    flows : dict
        Dictionary holding the flows that should be considered in
        constraint.
        Keys are (source, target) objects of the Flow. If no dictionary is
        given all flows containing the 'emission' attribute will be used.
    limit : numeric
        Absolute emission limit.
    Note
    -----
    Flow objects required an emission attribute!
    """

        if flows is None:
            flows = {}
            for (i, o) in om.flows:
                if hasattr(om.flows[i, o], 'emission'):
                    flows[(i, o)] = om.flows[i, o]

        else:
            for (i, o) in flows:
                if not hasattr(flows[i, o], 'emission'):
                    raise ValueError(('Flow with source: {0} and target: {1} '
                        'has no attribute '
                            'emission.').format(i.label, o.label))

        def emission_rule(m):
            return (sum(m.flow[inflow, outflow, t] * flows[inflow,
                outflow].emission
                for (inflow, outflow) in flows
                for t in m.TIMESTEPS) <= limit)
```

```
om.emission_limit = po.Constraint(rule=emission_rule)

return om
```

Another equation can be used to change a variable (*var*1), by multiplying this by a factor (*factor*1) (cf. Eq. 3.58) (oemof o.J.-c).

$$var1 \cdot factor1 = var2 \tag{3.58}$$

The parameter *factor*1 is a floating decimal point value set by the modeler. This parameter defines the relationship between the two numbers. This is described in the programming code in the following way:

```
def equate_variables(model, var1, var2, factor1=1, name=None):

    if name is None:
        name = '_'.join(["equate", str(var1), str(var2)])

    def equate_variables_rule(m):
        return var1 * factor1 == var2
    setattr(model, name, po.Constraint(rule=equate_variables_rule))
```

3.4   Mathematical Equations in the Module custom.py

To the module *custom.py* a few new classes have been added in Release v0.2.0. These were already listed in Chap. 2. Constraints are defined for the following two classes in this module:

- class *ElectricalLineBlock(SimpleBlock)*
- class *GenericCAESBlock(SimpleBlock)*

3.4.1   Class ElectricalLineBlock(SimpleBlock)

This module defines a constraint for the class *ElectricalLineBlock(SimpleBlock)* for the linear relation of a linear power flow (see Eq. 3.59) (oemof o.J.-c):

$$flow(n, o, t) = \frac{1}{reactrance(n, t)} \cdot (\text{voltage\_angle}(i(n), t) - \text{voltage\_angle}(o(n), t))$$

$$\tag{3.59}$$

This applies to:

$$\forall t \in TIMESTEPS \text{ and } \forall n \in ELECTRICAL\_LINES$$

The variable flow in Eq. 3.59 is calculated by means of the variable *reactance* and the variable voltage_angle. The parameter *reactance* is a floating decimal value, also called an array of float. This describes the reactance of a power line. The variable voltage_angle describes the voltage angle of a power line.

In the programming code this is formulated as *_voltage_angle_relation(block)* in the following excerpt:

```
def _voltage_angle_relation(block):
    for t in m.TIMESTEPS:
        for n in group:
            if O[n].slack is True:
                self.voltage_angle[O[n], t].value = 0
                self.voltage_angle[O[n], t].fix()
            try:
                lhs = m.flow[n, O[n], t]
                rhs = 1 / n.reactance[t] * (
                    self.voltage_angle[I[n], t] -
                    self.voltage_angle[O[n], t])
            except:
                raise ValueError("Error in constraint creation",
                                 "of node {}".format(n.label))
            block.electrical_flow.add((n, t), (lhs == rhs))
            # add constraint to set in-outflow equal
            block._equate_electrical_flows.add((n, t), (
                m.flow[n, O[n], t] == m.flow[I[n], n, t]))

self.electrical_flow = Constraint(group, noruleinit=True)

self._equate_electrical_flows = Constraint(group, noruleinit=True)

self.electrical_flow_build = BuildAction(rule=_voltage_angle_relation)
```

The method for generating linear constraints contains several functions, such as the function *def _voltage_angle_bounds(block, b, t)* and *def _voltage_angle_relation(block)*. But these are still in the development phase and cannot be applied yet.

3.4.2   Class GenericCAESBlock(SimpleBlock)

In Release v0.2.0 the modeling of a CAES plant is carried out by means of the class *GenericCAES*.

Following is a short excursus into the technology of compressed air energy storage, as this is a fairly new technology. Generally, storage power stations are known in the form of pumped-storage plants. Both types of storage plants, pumped-storage power plants and CAES plants provide interim storage for electricity, which can then be fed back to the grid when there is demand. Therefore, storage power plants can be used to balance demand between peak demand and low demand. In pumped-storage power stations, electricity is used to pump up water during low demand, mostly at night, in order to release energy during peak hours and thus provide energy during high demand to the grid that can be sold at a premium. Today, the main point of energy storage is not to balance high and low demand, but to store electricity from volatile energies, such as solar and wind energy and to make it available during times of low wind or low sunlight. Storage power stations create supply reliability when combined with volatile energy plants, as they can balance downtimes and uptimes of volatile plants. And they can also be used to cover peak demand.

But there is a difference between the two types of storage power. While pumped-storage power stations use water as the storage medium, compressed air energy storage stations use air as the storage medium. Furthermore, pumped-storage power stations make use of potential energy, by pumping water to a higher level. This potential energy can then be released by letting it flow downhill, driving turbines. In a compressed air energy storage plant, the excess energy is converted into pressure energy and stored. If the pressure is reduced the regained energy can be used to drive a turbine.

But how exactly does a compressed air energy storage plant work? If plants in the volatile energy sector produce more energy than can be fed back to the grid during high wind speeds and strong sunlight, this energy is used to operate compressors which compress air onto a higher pressure level. Pressures of up to 100 bars can be produced using this excess energy. The compressed air is stored in porous rock layers or underground caverns. The storage caverns must be large, airtight rooms. The compressed air is released and flows past turbines which convert the compressed air into kinetic energy, which is then converted back to electric energy by generators.

Compression and expansion of the air are associated with different temperatures. While compression causes a temperature rise, expansion cools the air to freezing temperatures. In order to prevent damage to the storage rooms, the compressed air must not get too hot. And for the turbines to run smoothly, the air must also not be too cold. There are two options for operating a CAES plant. To prevent the air from freezing, heat energy can be fed in from the outside. For this, compressed air is fed to a combustion chamber, where it is burned. Similarly, the compressed air must be cooled down. The cooling process in turn produces excess heat. Another option is to use the heat that is a by-product of the compression. But for this a thermic energy storage system is required, as the compression of the air and the cooling of the air happen at different points in time. For the second option, no additional heat source is required. For the first option, heat is normally produced by burning fossil fuels. Both operation options for CAES plants are described as a schematic diagram in Fig. 3.8.

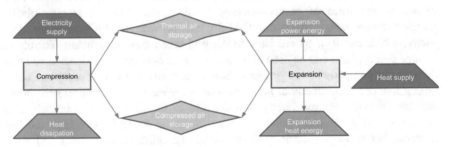

Fig. 3.8 Schematic of the operating principle of compressed air energy storage systems (CAES) (based on Kaldemeyer 2017)

A more detailed introduction to the operation of a CAES plant is forthcoming (Kaldemeyer 2017).

The class *GenericCAESBlock(SimpleBlock)* also requires several constraints to be defined. One constraint refers to the capacity of compressed air on the market and is formulated in the function *cmp_p_constr_rule(block, n, t)*:

```
def cmp_p_constr_rule(block, n, t):
    expr = 0
    expr += -self.cmp_p[n, t]
    expr += m.flow[list(n.electrical_input.keys())[0], n, t]
    return expr == 0
self.cmp_p_constr = Constraint(
    self.GENERICCAES, m.TIMESTEPS, rule=cmp_p_constr_rule)
```

Two constraints are introduced for the maximum capacity of compressed air and its dependence on the filling level of the cavern. These are formulated in the two functions *def cmp_p_max_constr_rule(block, n, t)* and *cmp_p_max_area_con-str_rule(block, n, t)*:

```
def cmp_p_max_constr_rule(block, n, t):
    if t != 0:
        return (self.cmp_p_max[n, t] ==
                n.params['cmp_p_max_m'] * self.cav_level[n, t-1] +
                n.params['cmp_p_max_b'])
    else:
        return (self.cmp_p_max[n, t] == n.params['cmp_p_max_b'])
```

```
self.cmp_p_max_constr = Constraint(
     self.GENERICCAES, m.TIMESTEPS, rule=cmp_p_max_constr_rule)

def cmp_p_max_area_constr_rule(block, n, t):
     return (self.cmp_p[n, t] <= self.cmp_p_max[n, t])
self.cmp_p_max_area_constr = Constraint(
     self.GENERICCAES, m.TIMESTEPS, rule=cmp_p_max_area_constr_rule)
```

Two more constraints formulated as (*cmp_st_p_min_constr_rule(block, n,)* and *cmp_st_p_max_constr_rule(block, n, t)*) allow to describe if a plant is in operation or is switched off:

```
def cmp_st_p_min_constr_rule(block, n, t):
     return (self.cmp_p[n, t] >= n.params['cmp_p_min'] * self.cmp_st[n, t])
self.cmp_st_p_min_constr = Constraint(
     self.GENERICCAES, m.TIMESTEPS, rule=cmp_st_p_min_constr_rule)

def cmp_st_p_max_constr_rule(block, n, t):
     return (self.cmp_p[n, t] <=
             (n.params['cmp_p_max_m'] * n.params['cav_level_max'] +
             n.params['cmp_p_max_b']) * self.cmp_st[n, t])
self.cmp_st_p_max_constr = Constraint(
     self.GENERICCAES, m.TIMESTEPS, rule=cmp_st_p_max_constr_rule)
```

The heat flow from the compression is described by the constraint (*cmp_q_out_constr_rule(block, n, t)*):

```
def cmp_q_out_constr_rule(block, n, t):
     return (self.cmp_q_out_sum[n, t] ==
             n.params['cmp_q_out_m'] * self.cmp_p[n, t] +
             n.params['cmp_q_out_b'] * self.cmp_st[n, t])
self.cmp_q_out_constr = Constraint(
    self.GENERICCAES, m.TIMESTEPS, rule=cmp_q_out_constr_rule)
```

For a single heat outflow from compression, there is also a constraint (*cmp_q_out_sum_constr_rule(block, n, t)*):

```
def cmp_q_out_sum_constr_rule(block, n, t):
     return (self.cmp_q_out_sum[n, t] == self.cmp_q_waste[n, t] +
         self.tes_e_in[n, t])
self.cmp_q_out_sum_constr = Constraint(
    self.GENERICCAES, m.TIMESTEPS, rule=cmp_q_out_sum_constr_rule)
```

Another constraint (*cmp_q_out_shr_constr_rule(block, n, t)*) looks at the interdependence between the variable of heat loss *self.cmp_q_waste[n, t]* and the variable of the heat inflow from thermal energy storage (TES) *self.tes_e_in[n, t]*:

```
def cmp_q_out_shr_constr_rule(block, n, t):
    return (self.cmp_q_waste[n, t] * n.params['cmp_q_tes_share'] ==
        self.tes_e_in[n, t] * (1 - n.params['cmp_q_tes_share']))
self.cmp_q_out_shr_constr = Constraint(
    self.GENERICCAES, m.TIMESTEPS, rule=cmp_q_out_shr_constr_rule)
```

For the expansion the capacity on the market is described via the function *exp_p_constr_rule(block, n, t)*:

```
def exp_p_constr_rule(block, n, t):
    expr = 0
    expr += -self.exp_p[n, t]
    expr += m.flow[n, list(n.electrical_output.keys())[0], t]
    return expr == 0
self.exp_p_constr = Constraint(
    self.GENERICCAES, m.TIMESTEPS, rule=exp_p_constr_rule)
```

Another constraint can be defined for the expansion of the maximum capacity in relation to the cavern filling level via the function *exp_p_max_constr_rule(block, n, t)*:

```
def exp_p_max_constr_rule(block, n, t):
    if t != 0:
        return (self.exp_p_max[n, t] ==
            n.params['exp_p_max_m'] * self.cav_level[n, t-1] +
            n.params['exp_p_max_b'])
    else:
        return (self.exp_p_max[n, t] == n.params['exp_p_max_b'])
self.exp_p_max_constr = Constraint(
    self.GENERICCAES, m.TIMESTEPS, rule=exp_p_max_constr_rule)

def exp_p_max_area_constr_rule(block, n, t):
    return (self.exp_p[n, t] <= self.exp_p_max[n, t])
self.exp_p_max_area_constr = Constraint(
    self.GENERICCAES, m.TIMESTEPS, rule=exp_p_max_area_constr_rule)
```

For the expansion the operation mode (on/off) must be read via the function *exp_st_p_min_constr_rule(block, n, t)*:

```
def exp_st_p_min_constr_rule(block, n, t):
    return (
        self.exp_p[n, t] >= n.params['exp_p_min'] * self.exp_st[n, t])
self.exp_st_p_min_constr = Constraint(
    self.GENERICCAES, m.TIMESTEPS, rule=exp_st_p_min_constr_rule)


def exp_st_p_max_constr_rule(block, n, t):
    return (self.exp_p[n, t] <=
        (n.params['exp_p_max_m'] * n.params['cav_level_max'] +
        n.params['exp_p_max_b']) * self.exp_st[n, t])
self.exp_st_p_max_constr = Constraint(
    self.GENERICCAES, m.TIMESTEPS, rule=exp_st_p_max_constr_rule)
```

For the incoming heat flow during expansion the function *exp_q_in_constr_rule (block, n, t)* applies:

```
def exp_q_in_constr_rule(block, n, t):
    return (self.exp_q_in_sum[n, t] ==
        n.params['exp_q_in_m'] * self.exp_p[n, t] +
        n.params['exp_q_in_b'] * self.exp_st[n, t])
self.exp_q_in_constr = Constraint(
    self.GENERICCAES, m.TIMESTEPS, rule=exp_q_in_constr_rule)
```

For the fuel allocation during expansion the following constraint (*exp_q_fuel_constr_rule(block, n, t)*) can be formulated:

```
def exp_q_fuel_constr_rule(block, n, t):
    expr = 0
    expr += -self.exp_q_fuel_in[n, t]
    expr += m.flow[list(n.fuel_input.keys())[0], n, t]
    return expr == 0
self.exp_q_fuel_constr = Constraint(
    self.GENERICCAES, m.TIMESTEPS, rule=exp_q_fuel_constr_rule)
```

For the expansion there is also a single heat flow, which is described by the function *exp_q_in_sum_constr_rule(block, n, t)*:

```
def exp_q_in_sum_constr_rule(block, n, t):
    return (self.exp_q_in_sum[n, t] == self.exp_q_fuel_in[n, t] +
        self.tes_e_out[n, t] + self.exp_q_add_in[n, t])
self.exp_q_in_sum_constr = Constraint(
    self.GENERICCAES, m.TIMESTEPS, rule=exp_q_in_sum_constr_rule)
```

The constraint describing the relation between the input heat flows during the expansion is written as (*exp_q_in_shr_constr_rule(block, n, t)*):

```
def exp_q_in_shr_constr_rule(block, n, t):
    return (n.params['exp_q_tes_share'] * self.exp_q_fuel_in[n, t] ==
        (1 - n.params['exp_q_tes_share']) *
        (self.exp_q_add_in[n, t] + self.tes_e_out[n, t]))
self.exp_q_in_shr_constr = Constraint(
    self.GENERICCAES, m.TIMESTEPS, rule=exp_q_in_shr_constr_rule)
```

Let us now look at the cavern. One constraint describes the energy inflow *cav_e_in_constr_rule(block, n, t)*:

```
def cav_e_in_constr_rule(block, n, t):
    return (self.cav_e_in[n, t] ==
        n.params['cav_e_in_m'] * self.cmp_p[n, t] +
        n.params['cav_e_in_b'])
self.cav_e_in_constr = Constraint(
    self.GENERICCAES, m.TIMESTEPS, rule=cav_e_in_constr_rule)
```

There is also a constraint describing the energy outflow from the cavern *cav_e_out_constr_rule(block, n, t)*:

```
def cav_e_out_constr_rule(block, n, t):
    return (self.cav_e_out[n, t] ==
        n.params['cav_e_out_m'] * self.exp_p[n, t] +
        n.params['cav_e_out_b'])
self.cav_e_out_constr = Constraint(
    self.GENERICCAES, m.TIMESTEPS, rule=cav_e_out_constr_rule)
```

For the storage cavern, there is also the constraint that the energy flow must be balanced overall. This is described by the function *cav_eta_constr_rule(block, n, t)*:

```
def cav_eta_constr_rule(block, n, t):
    if t != 0:
        return (n.params['cav_eta_temp'] * self.cav_level[n, t] ==
            self.cav_level[n, t-1] + m.timeincrement[t] *
            (self.cav_e_in[n, t] - self.cav_e_out[n, t]))
    else:
        return (n.params['cav_eta_temp'] * self.cav_level[n, t] ==
            m.timeincrement[t] *
            (self.cav_e_in[n, t] - self.cav_e_out[n, t]))
self.cav_eta_constr = Constraint(
    self.GENERICCAES, m.TIMESTEPS, rule=cav_eta_constr_rule)
```

An upper bound is defined for the storage cavern via the function *cav_ub_ constr_rule(block, n, t)*:

```
def cav_ub_constr_rule(block, n, t):
    return (self.cav_level[n, t] <= n.params['cav_level_max'])
self.cav_ub_constr = Constraint(
    self.GENERICCAES, m.TIMESTEPS, rule=cav_ub_constr_rule)
```

A balanced energy flow is also required for the thermal energy storage (TES) described via the function *tes_eta_constr_rule(block, n, t)*:

```
def tes_eta_constr_rule(block, n, t):
    if t != 0:
        return (self.tes_level[n, t] ==
            self.tes_level[n, t-1] + m.timeincrement[t] *
        (self.tes_e_in[n, t] - self.tes_e_out[n, t]))
    else:
        return (self.tes_level[n, t] ==
            m.timeincrement[t] *
            (self.tes_e_in[n, t] - self.tes_e_out[n, t]))
self.tes_eta_constr = Constraint(
    self.GENERICCAES, m.TIMESTEPS, rule=tes_eta_constr_rule)
```

By means of an upper bound for energy storage (TES) overload can be prevented. This upper bound is introduced into the model via the function *tes_ub_ constr_rule(block, n, t)*:

```
def tes_ub_constr_rule(block, n, t):
    return (self.tes_level[n, t] <= n.params['tes_level_max'])
self.tes_ub_constr = Constraint(
    self.GENERICCAES, m.TIMESTEPS, rule=tes_ub_constr_rule)
```

3.5   Methods Provided in the Module models.py

The module *models.pv* does not contain mathematical modules but provides some methods. Some methods are listed here for the class *BaseModel(po. ConcreteModel)*. These methods are relevant for developing models. The methods are listed in Table 3.1.

Table 3.1 Methods in the module *models.py* of the class *BaseModel(po.ConcreteModel* (oemof 2014a)

Method	Description
def _add_objective(self, sense=po. minimize, update=False)	Method to sum up all objective expressions from the child blocks that have been created. This method looks for *_objective_expression* attribute in the block definition and will call this method to add their return value to the objective function.
def receive_duals(self)	Method sets solver suffix to extract information about dual variables from solver. Shadow prices (duals) and reduced costs (rc) are set as attributes of the model.
def relax_problem(self)	Relaxes integer variables to reals of optimization model self.
def results(self)	Returns a nested dictionary of the results of this optimization.
solve(solver='cbc', solver_io='lp', **kwargs)	Takes care of communication with solver to solve the model.

More information can be found in GitHub (2017c)

References

Amazon Web Services (AWS): Unterstützte Elastic Beanstalk-Plattformen. In: AWS: AWS Elastic Beanstalk: Entwicklerhandbuch (2018). https://docs.aws.amazon.com/de_de/elasticbeanstalk/latest/dg/concepts.platforms.html. Accessed on 23 Feb 2018

Chen, P.P.-S.: The entity-relationship model—toward a unified view of data. In: ACM Transactions on Database Systems (TODS)—Special Issue: Papers from the International Conference on Very Large Data Bases: 22–24 Sept 1975, Framingham, MA. vol. 1, no. 1, 03-1976, pp. 9–36

Domschke, W., et al.: Einführung in Operations Research, 9. Auflage. Springer Gabler, Berlin (2015)

GitHub Inc.: oemof/oemof/solph/ (2017a). https://github.com/oemof/oemof/tree/dev/oemof/solph. Accessed on 18 Dec 2017

GitHub Inc.: oemof/oemof/solph/components.py. (2017b). https://github.com/oemof/oemof/blob/dev/oemof/solph/components.py. Accessed on 18 Dec 2017

GitHub Inc.: oemof/oemof/solph/models.py (2017c). https://github.com/oemof/oemof/blob/dev/oemof/solph/models.py. Accessed on 18 Dec 2017

GitHub Inc.: oemof/oemof/solph/blocks.py (2017d). https://github.com/oemof/oemof/blob/dev/oemof/solph/blocks.py. Accessed on 20 Dec 2017

GitHub Inc.: oemof/oemof/ (2018a). https://github.com/oemof/oemof/tree/dev/oemof, Accessed on 26 Feb 2018

GitHub Inc.: oemof/oemof (2018b). https://github.com/oemof/oemof. Accessed on 26 Feb 2018

Kaldemeyer, C. et al.: A Generic Formulation of Compressed Air Energy Storage as Mixed Integer Linear Program – Unit Commitment of Specific Technical Concepts in Arbitrary Market Environments Materials Today. Selection and/or Peer-review under responsibility of 5th International Conference on Nanomaterials and Advanced Energy Storage Systems

Mistakidis, E.S., et al.: Nonconvex Optimization in Mechanics: Algorithms, Heuristics and Engineering Applications by the F.E.M. Springer, Berlin (2011)

Mollenhauer, E., et al.: Evaluation of an energy- and exergy-based generic modelling approach of combined heat and power plants. Int. J. Energy Environ. Eng. **7**(2), 167–176 (2016) (Springer). https://link.springer.com/content/pdf/10.1007%2Fs40095-016-0204-6.pdf. Accessed 03 Jan 2018

oemof-developer-group: oemof package (2014a). http://oemof.readthedocs.io/en/latest/api/oemof.solph.html#oemof.solph.options.NonConvex. Accessed on 04 Jan 2018

oemof-developer-group: Meta description (2014b). http://www.pythonhosted.org/oemof_base/meta_description.html. Accessed on 04 Jan 2018

oemof-developer-group: oemof.solph package (2014c). http://oemof.readthedocs.io/en/latest/api/oemof.solph.html#module-oemof.solph.blocks. Accessed on 09 Jan 2018

oemof-developer-group: Source code for oemof.solph.blocks (2014d). http://oemof.readthedocs.io/en/latest/_modules/oemof/solph/blocks.html#Flow. Accessed on 09 Jan 2018

oemof-developer-group: oemof.solph package, o.J.-c. http://oemof.readthedocs.io/en/latest/api/oemof.solph.html#module-oemof.solph.components . Accessed 22 Feb 2018

Pyomo: Pyomo online Documentation 3.5 (2014). https://software.sandia.gov/.Downloads/pub/coopr/CooprGettingStarted.html#_buildaction. 21 Feb 2018

Pyomo: Pyomo Documentation : Release 5.1 (2014). https://media.readthedocs.org/pdf/pyomo/expr_dev/pyomo.pdf. 21 Feb 2018

Salgado, F., et al.: Short-term operation planning on cogeneration systems: a survey. Electric Power Syst. Res. **78**(5), 835–848 (2008) (Elsevier). https://doi.org/10.1016/j.epsr.2007.06.001. Accessed on 10 Jan 2018

Werners, B.: Grundlagen des Operations Research: Mit Aufgaben und Lösungen, 3 Auflage. Springer Gabler, Berlin (2013)

Windisch, H.: Thermodynamik : Ein Lehrbuch für Ingenieure, 5. Auflage. Oldenburg Wissenschaftsverlag GmbH, München (2014)

Chapter 4
Modeling in OEMOF

The OEMOF framework is highly flexible and can be used for solving very different kinds of problems and answering different types of questions. OEMOF allows you to create both dispatch models and investment models. Key factors in creating a successful model are detailed knowledge of the region, the country or the area that is modeled as well as a thorough understanding of the energy supply and energy technology aspects. Following on from the previous chapter's description of the mathematical basis of the program, this chapter will describe how the different modules, packages, and libraries interact. But before the modeling can start, datasets describing the technical and economical situation are required. Thus procurement of data is the prerequisite to successful modeling. No optimization model can be created without successful data gathering. However, reliable and relevant data are not always easy to find. Another difficulty is keeping track of the code that has already been written for your model. This is achieved by means of a surface that acts as an interface between model and solver. This book introduces the program system Spyder as an example for such an interface.

4.1 Working with the OEMOF Framework

Chapter 3 focused on the mathematical basis of the objective function and of some key constraints which will form part of the model created. This chapter will focus on the role that the different modules, packages and libraries will have for the model. The equations and inequations are formulated in methods and functions that are written into the code and then executed during the computation of an optimization problem. But OEMOF also contains some methods and functions that support the main modeling tasks, for example grouping and sorting elements. Even though those methods and functions are not a part of the actual optimization model, these are also introduced and described briefly in this chapter. Some excerpts from

© Springer International Publishing AG, part of Springer Nature 2019
J. Nagel, *Optimization of Energy Supply Systems*, Lecture Notes in Energy 69,
https://doi.org/10.1007/978-3-319-96355-6_4

the programming code will be given here in order to give an illustration of how they are used in the user applications.

In OEMOF programming and modeling are viewed as two different applications. Modelers use OEMOF as a tool. They do not necessarily require detailed knowledge of the internal structures of the programming code. Nevertheless, modelers who do want to go into creating classes, methods or functions can find the relevant information for this in OEMOF. This chapter is designed to give an overview of the existing program structures to users who are interested in working on further developing the OEMOF programs. A more detailed insight into the Python programming language is given in Sect. 4.1.1 as well as some other sections of this chapter.

4.1.1   A Deeper Insight into Programming in Python

There is one more type of element in Python that we have not yet discussed. This is a so-called *decorator*. The most commonly used decorators are (Stack Overflow o.J.):

– *@property*
– *@classmethod*
– *@staticmethod.*

Decorators are also part of the OEMOF set of programs. A key syntax character is the @ symbol, which can be used for class decorators, function decorators, or method decorators (Stack Overflow o.J.). Decorators indicate that something new is happening, that some settings have been changed (Python 2003).

A decorator is a software design pattern (Python 2016). These are tried and tested templates for recurring design problems. Design problems can occur both in software development and in software architecture. Decorators are attached to individual objects of a class, either as dynamic or static adjuncts, but do not affect the code of other objects in the same class that do not have the decorator. Attaching a decorator to an object changes the effect of methods, functions, or classes on this object. As a programming element decorators are themselves functions.

In contrast to other decorators, Python decorators are a particularly versatile tool for modifying Python syntax (Python 2016). Using decorators is a very convenient tool for altering or modifying methods or functions. Using decorators also increases the readability of the source code.

In the OEMOF programming code, decorators are introduced immediately before the definition of a method or function. This can be seen in the excerpt below:

```
@staticmethod
    def regroup(entity, groups, groupings):
    pass
```

The following section will introduce some frequently Python decorators in OEMOF.

Decorator *@property*

The decorator *@property* is defined as part of a class (Python 2018a):

```
class property(object):
      def __init__(fget=None, fset=None, fdel=None, doc=None):
      pass
```

where:

– *fget*
  is a function that fetches a value for an attribute and can read that value.
– *fset*
  This function sets a value for an attribute and as a result changes the previous value.
– *fdel*
  This function deletes the value of an attribute.
– *doc*
  This creates a docstring.

The output is a property attribute.

Excursus: What is a docstring? As we have already seen in previous chapters, we often have text elements below the function framed by quotation marks at the beginning and at the end of the text. This syntax indicates to Python that this is a documentation string (Python For Beginners o.J.). This character string is to describe the role of the function or method that follows. According to the Python manual, this should be done by using three quotation marks in a row at each end of the documentation string. This information can then be used to generate documentation by means of a documentation generator. The OEMOF code excerpt below gives and example for this:

```
""" Add :class:'nodes <oemof.network.Node>' to this energy system.
" " "
```

But now, let us go back to the decorator. The decorator *property* is mainly used to manage private attributes. To do this, it must be capable of reading and modifying an attribute. But as private attributes are encapsulated, they cannot be easily accessed. Normally, two methods must be written into the code to provide access, one method for reading the attribute value (e.g., *getX()*) and another method for altering the value (e.g., *setX()*). This is a query method. The first method *getX()* is called the getter and the second method *setX()* is called the setter (Klein o.J.-a). Altering an existing attribute is a fairly typical application for a decorator (Python 2018a). An example of this can be seen in the code excerpt below. In this sequence, the class *S* has the attribute *x*:

```
class S:
    def __init__(self):
        self._x = x

    def getx(self):
        return self._x

    def setx(self, x):
        self._x = x

    def delx(self):
        del self._x
```

This becomes more specific in the next excerpt:

```
>>> from s import S
>>> a = S(5)
>>> a.getX()
5
>>> a.setX(10)
>>> b = a.getX() * 0.21
>>> b
2.100000000000000
>>>
```

The following line of code is the one that is to be replaced by the decorator:

```
b = a.getX() * 0.21
```

Having to write this line into every single sequence makes the code less clear and requires more writing. In order to simplify the source code, we add the decorator *property*. Again, we use class S with the attribute x:

```
class S:
    def __init__(self):
        self._x = x

    def getx(self):
        return self._x

    def setx(self, x):
        self._x = x
```

```
def delx(self):
    del self._x
```

```
x = property(getx, setx, delx, "I'm the 'x' property.")
```

Next, we initialize an instance *s* of the class *S*. Normally *return S._x* would call the getter. The next sequence *S._x=x* would then trigger the setter. The sequence starting with the instruction *del S._x* would delete the content of *x* and generate the string *I'm the 'x' property*.

Using the decorator allows for the simplified code sequence below (Klein o.J.-d):

```
>>> from s import S
>>> a = S(10)
>>> b = P(20)
>>> c = a.x + b.x
>>> c
30
>>>
```

Let us look at another specific example from the energy sector. First, we introduce the class *LowVoltageGrid*. Now the attribute *x* becomes the attribute *voltage* which is given a specific value. Next the decorator *property* is called and the attribute *voltage* (100,000) together with the docstring *I'm the 'x' property* is overwritten and replaced by *Get the current voltage*. This is illustrated by the source code excerpt below (Python 2018a):

```
class LowVoltageGrid:
    def __init__(self):
        self._voltage = 100000

    @property
    def voltage(self):
        """Get the current voltage."""
        return self._voltage
```

Decorator @classmethod

This decorator transforms a method into a class method (Python 2018a). A class method always carries the class name as its first implicit argument. This can be implemented in the following way in the source code. For this example, we are using the class *grid*:

```
class grid:
    @classmethod
    def f(cls, arg1, arg2, ...): ...
```

More information about this can be found in (Python 2018a).

Decorator *@staticmethod*

By means of this decorator a method is transformed into a static method. The advantage of a static method is that this type of method does not have to start out by processing an implicit argument. Static methods do not require an object instance, but can be called directly via the class. For this reason, a static method does not require the user to call the object instance *self* as first command, as this object instance does not yet exist when the class is first called. But this also means that static methods cannot access the attributes of an object. Further information about this can be found in (Python 2018a).

Decorator *@nodes.setter*

This is another type of decorator that is available in OEMOF. The method *setter*, which is a so-called mutator method is responsible for changing data (Klein o.J.-a). The following method, which is used as a setter, is decorated by *@nodes.setter*. *Nodes* is here the name of the function. If the decorator was added to a function by the name of *p*, the method would be called by the command *@p.setter*.

Built-in functions

Decorators are one of the so-called built-in functions. The Python interpreter offers several functions and types that have been built into it. That means that they are always available. The built-in functions are listed in Table 4.1.

Table 4.1 Python built-in functions (Python 2018a)

Built-in functions				
abs()	dict()	help()	min()	setattr()
all()	dir()	hex()	next()	slice()
any()	divmod()	id()	object()	sorted()
ascii()	enumerate()	input()	oct()	staticmethod()
bin()	eval()	int()	open()	str()
bool()	exec()	isinstance()	ord()	sum()
bytearray()	filter()	issubclass()	pow()	super()
bytes()	float()	iter()	print()	tuple()
callable()	format()	len()	property()	type()
chr()	frozenset()	list()	range()	vars()
classmethod()	getattr()	locals()	repr()	zip()
compile()	globals()	map()	reversed()	__import__()
complex()	hasattr()	max()	round()	
delattr()	hash()	memoryview()	set()	

One function that comes up quite often in the OEMOF set of programs is *hasattr* *(object, name)*. The arguments of this function are an object and a string. The result of this term is *True*, if the string corresponds to the name of the attribute of an object. The output is *False* if that is not the case.

Magic methods

There are so-called magic methods in Python. These all have a double underscore __ at the beginning and at the end of the method. These methods are carried out automatically by Python in the background, as if by magic. One example for such a magic method is the method *__call__()*. By means of this method an arbitrary instance of a class can be called (Python 2018b). In order to do this, this method must be introduced in the relevant class. This method is useful for querying the arguments of an object. This can be executed in the following way:

```
# An instance of 'HighVoltage' of the class 'object' is created and then
# the arguments of this instance are queried.
class HighVoltage(object):
    def __call__(self, voltage, frequence):
        print voltage
        print frequence

t=Test
t(1000, 50)
```

The expected output would be the following:

```
(1000, 50)
```

In order to call this function, the generated object is represented in the source code via the method *__call__()* (Python 2018b):

```
HighVoltageBerlin.__call__(self[, voltage, frequence])
```

Now the object can be called and the values returned. Understanding these methods and functions helps us to interpret more excerpts from the source code.

4.1.2   Oemof-Energy_System

The module *energy_system.py* is a key module in the OEMOF framework. It plays a vital role in the OEMOF structure. This module defines the base class *EnergySystem* (GitHub 2017a). The class *EnergySystem* contains a number of relevant parameters and attributes, which were introduced in Sect. 2.4.1. The first

method ___init___ introduces the attributes for the instances of the class
EnergySystem.

```
def __init__(self, **kwargs):
    for attribute in ['entities']:
        setattr(self, attribute, kwargs.get(attribute, []))

    self._groups = {}
    self._groupings = ([BY_UID] +
            [g if isinstance(g, Grouping) else Nodes(g)
                for g in kwargs.get('groupings', [])])
    for e in self.entities:
        for g in self._groupings:
            g(e, self.groups)
    self.results = kwargs.get('results')
    self.timeindex = kwargs.get('timeindex',
                        pd.date_range(start=pd.to_datetime('today'),
                        periods=1, freq='H'))
```

Next we will look at the decorator *@staticmethod* used for the following
method:

– *def _regroup(entity, groups, groupings)*
  This method allows for a regrouping of the entities within a group.

  Further methods follow:

– *def _add(self, entity)*
  This method can be used to add entities.
– *def add(self, *nodes)*
  This method adds additional *nodes*.

These methods are represented in the following excerpt from the source code:

```
@staticmethod
def _regroup(entity, groups, groupings):
    for g in groupings:
        g(entity, groups)
    return groups


def _add(self, entity):
    self.entities.append(entity)
    self._groups = partial(self._regroup, entity, self.groups,
                    self._groupings)
```

```
def add(self, *nodes):
    """ Add :class:'nodes <oemof.network.Node>' to this energy system.
    """
    for n in nodes:
        self._add(n)
```

The decorator @*property* appears twice to change attributes.

```
@property
def groups(self):
# The next three lines mean: While the object self._groups
# is callable this object is called by self._groups()
# and the output self._groups is reassigned and tested as to its
# callability. As soon as self._groups is no longer callable
# its value is returned as the attribute value of groups.
    while callable(self._groups):
        self._groups = self._groups()
    return self._groups
```

The second method is:

```
@property
def nodes(self):
    return self.entities
```

Below the line with the decorator @*nodes.setter* the method *nodes* is executed:

- *def nodes(self, value)*
  Here values are passed to the entities:

Next a few more methods:

- *def flows(self)*
  This method is used to divide flows into inputs and outputs.
- *def dump(self, dpath=None, filename=None)*
  The method *dump()* serializes the hierarchy of an object.
- *def restore(self, dpath=None, filename=None)*
  An instance of an energy system is returned.

These can all be seen in the following excerpt:

```
@nodes.setter
def nodes(self, value):
    self.entities = value
```

```python
def flows(self):
    return {(source, target): source.outputs[target]
        for source in self.nodes
        for target in source.outputs}


def dump(self, dpath=None, filename=None):
    """ Dump an EnergySystem instance.
    """
    if dpath is None:
        bpath = os.path.join(os.path.expanduser("~"), '.oemof')
        if not os.path.isdir(bpath):
            os.mkdir(bpath)
        dpath = os.path.join(bpath, 'dumps')
        if not os.path.isdir(dpath):
            os.mkdir(dpath)

    if filename is None:
        filename = 'es_dump.oemof'
# The module "pickle" implements a binary protocol in order to
# serialize object structures in Python [Python 2018c].
# In computer science, serializing is defined as the transformation of
# structured data into a sequential presentation.
# This can be used for increasing persistence of object data, i.e.
# when objects and their data are to be made available across a certain
# time period. And serializing is relevant if a widespread software network
# is used and objects are to be transferred across it. Via this process
# a Python object hierarchy is transformed into a byte stream structure
# The dump() function of the module pickle serializes a dictionary object
# in an open file. Via self._dict_ the attribute dictionary of the object
# is passed. The following command means: Pickle the 'data' dictionary
# by making use of the available protocol [Python 2018c]. The command
# 'open()' generates a file object in binary code. The abbreviation 'wb'
# stands for: 'write binary'.
# NB: The term 'pickle' is used to refer to serialization

    pickle.dump(self.__dict__, open(os.path.join(dpath, filename),
                'wb'))


# 'msg' is a messaging library. This library can be used to create
# a live chat or instant messenger service [Python o.J.].

    msg = ('Attributes dumped to: {0}'.format(os.path.join(
            dpath, filename)))
```

```
logging.debug(msg)
return msg

def restore(self, dpath=None, filename=None):
r""" Restore an EnergySystem instance.
" " "
logging.info(
  "Restoring attributes will overwrite existing attributes.")
if dpath is None:
  dpath = os.path.join(os.path.expanduser("~"), '.oemof',
                     'dumps')

if filename is None:
  filename = 'es_dump.oemof'

self.__dict__ = pickle.load(open(os.path.join(dpath, filename),
                     "rb"))
msg = ('Attributes restored from: {0}'.format(os.path.join(
    dpath, filename)))
logging.debug(msg)
return msg
```

The OEMOF system of programs groups all elements of the system according to their unique identifier UID. This can be seen in the following source code excerpt (GitHub 2017a):

```
from oemof.network import Entity
from oemof.network import Bus, Sink
es = EnergySystem()
bus = Bus(label='electricity')
es.add(bus)
bus is es.groups['electricity']
    True
```

The first line imports the base class *Entity*. The second line imports additional required nodes, in this case *Bus* and *Sink*. So that we do not have to write the full name of the class *EnergySystem*, the abbreviation *es* is assigned. Next the class *Bus* receives the label *electricity* and the information is given that *Bus* is called as *bus* in the source code. The next line defines *bus* as a subclass of *EnergySystem*. Now the classes are grouped by the label *electricity*. If all these conditions apply, the value*True* is to be returned. This is a very simple format for carrying out grouping.

4.1.3   Oemof-Groupings

The module *groupings* is an OEMOF module that runs in the background. Modelers do not require to actively use it as part of their model. But it is relevant for developers. We have already mentioned groupings or group a few times. But before the methods and functions of this module in OEMOF are described, we will briefly talk about the practical value of this module.

How can this module be used? For example, all gas-powered transformers can be grouped together. By grouping them together, constraints, variables and object characteristics can be defined for the whole group. These constraints, variables and object characteristics will then be available to each element of the group. Another option for using grouping is to return nodes with a predefined attribute. For example, the attribute could be ‚all buses that are balanced'. This means that each node of the energy system has to be analyzed and then a key must be returned depending on the attribute. This can be written in the source code like this:

```
def constraint_grouping(node):
    if isinstance(node, Bus) and node.balanced:
        return blocks.Bus
    if isinstance(node, Transformer):
        return blocks.Transformer
GROUPINGS = [constraint_grouping]
```

This function can then be transferred to a list format. Two lists will be generated, one list contains all transformers, the other list contains all balanced buses. These groups are then stored in a dictionary.

The grouping function and the lists it produces as an output format can also be useful for returning the results after computing the optimization function for an energy system; e.g., for looking at specific entities of the model. This requires a method for accessing a dictionary and for returning the contents.

Let us now look specifically at how to use the module *groupings* in OEMOF. This module defines the methods for the classes *Grouping*, *Nodes(Grouping)*, *Flows(Nodes)* and *FlowsWithNodes(Nodes)*.

For the class *Grouping* first the method *__init__* is introduced:

```
def __init__(self, key=None, constant_key=None, filter=None, **kwargs):
    if key and constant_key:
        raise TypeError(
                "Grouping arguments 'key' and 'constant_key' are " +
                    " mutually exclusive.")
    if constant_key:
        self.key = lambda _: constant_key
    elif key:
```

```
        self.key = key
    else:
        raise TypeError(
                "Grouping constructor missing required argument: " +
                "one of 'key' or 'constant_key'.")
    self.filter = filter
    for kw in ["value", "merge", "filter"]:
        if kw in kwargs:
            setattr(self, kw, kwargs[kw])
```

The next method provides a key under which the group is stored:

```
def key(self, e):
```

Next the group is generated with the entities it contains:

```
def value(self, e):
    """ Generate the group obtained from :obj:'e'.
    This methd returns the actual group obtained from :obj:'e'. Like
    :meth:'key <Grouping.key>', it is called for every :obj:'e' in the
    energy system. If there is no group stored under :meth:'key(e)
    <Grouping.key>', :obj:'groups[key(e)]' is set to :meth:'value(e)
    <Grouping.value>'. Otherwise :meth:'merge(value(e), groups[key(e)])
    <Grouping.merge>' is called.
    The default returns the :class:'entity <oemof.core.network.Entity>'
    itself.
    """
    return e
```

If a new group is generated where a previous group exists already, these can be merged by the method *def merge(self, new, old)*:

```
def merge(self, new, old):
    """ Merge a known :obj:'old' group with a :obj:'new' one.
    This method is called if there is already a value stored under
    :obj:'group[key(e)]'. In that case, :meth:'merge(value(e),
    group[key(e)]) <Grouping.merge>' is called and should return the new
    group to store under :meth:'key(e) <Grouping.key>'.

    The default behaviour is to raise an error if :obj:'new' and
    :obj:'old' are not identical.
    """
```

```
if old is new:
    return old
raise ValueError("\nGrouping \n  " +
                 "{}:{}\nand\n {}:{}\ncollides.\n".format(
                 id(old), old, id(new), new) +
                 "Possibly duplicate uids/labels?")
```

The method *def filter(self, group)* allows users to set a filter before storing the entities of a group. This means that only those entities are included in the group that were selected by the filter:

```
def filter(self, group):
    ''''''
    :func:'Filter <builtins.filter>' the group returned by :meth:'value'
    before storing it.
    Should return a boolean value. If the :obj:'group' returned by
    :meth:'value' is :class:'iterable <collections.abc.Iterable>', this
    function is used (via Python's :func:'builtin filter
    <builtins.filter>') to select the values which should be retained in
    :obj:'group'. If :obj:'group' is not :class:'iterable
    <collections.abc.Iterable>', it is simply called on :obj:'group'
    Itself and the return value decides whether :obj:'group' is stored
    (:obj:'True') or not (:obj:'False').
```

Here the method *def __call__(self, e, d)* is introduced:

```
def __call__(self, e, d):
    k = self.key(e) if callable(self.key) else self.key
    if k is None:
        return
    v = self.value(e)
    if isinstance(v, MuMa):
        for k in list(filterfalse(self.filter, v)):
            v.pop(k)
    elif isinstance(v, Mapping):
        v = type(v)((k, v[k]) for k in v if self.filter(k))
    elif isinstance(v, Iterable):
        v = type(v)(filter(self.filter, v))
    elif self.filter and not self.filter(v):
        return
    if not v:
        return
```

```
for group in (k if (isinstance(k, Iterable) and not
                    isinstance(k, Hashable))
              else [k]):
    d[group] = (self.merge(v, d[group]) if group in d else v)
```

Once the class *Grouping* has been initialized, the class *Nodes(Grouping)* can be defined in the programming code. This class inherits all attributes from the class *Grouping*. And for this subclass the method *def value(self, e)* must again be formulated, this time for all nodes.

```
def value(self, e):
    """ 
    Returns a :class:'set' containing only :obj:'e', so groups are
    :class:'sets <set>' of :class:'node <oemof.network.Node>'.
    """ 
        return {e}
```

This class is also given a method for merging an old group with a new group:

```
def merge(self, new, old):
    """ 
    :meth:'Updates <set.update>' :obj:'old' to be the union of :obj:'old'
    and :obj:'new'.
    """ 
    return old.union(new)
```

The class *Flow* contained in this module inherits from the previous class *Nodes*. The method *def value(self, flows)* contains the argument 'flows'. This can be used to return the values for the flows:

```
def value(self, flows):
    '''''' 
    Returns a :class:'set' containing only :obj:'flows', so groups are
    :class:'sets <set>' of flows.
    '''''' 
    return set(flows)
```

We can see that the subclasses always contain the same methods as the parent class, but each subclass has its own filter. Next we will look at the method *def __call__(self, n, d)*.

```
def __call__(self, n, d):
    flows = set(chain(n.outputs.values(), n.inputs.values()))
    super().__call__(flows, d)
```

The class *FlowsWithNodes(Nodes)* inherits all attributes and methods from the parent class *Nodes*. But the filter ensures that only flows are returned that are connected to a node. The result of this filter selection is tuples. For these tuples the method *def value(self, tuples)* is executed again.

```
def value(self, tuples):
    " " "
    Returns a :class:'set' containing only :obj:'tuples', so groups are
    :class:'sets <set>' of :obj:'tuples'.
    " " "
    return set(tuples)
```

The method *def __call__(self, n, d)* is also executed for the tuples:

```
def __call__(self, n, d):
    tuples = set(chain(
                ((n, t, f) for (t, f) in n.outputs.items()),
                ((s, n, f) for (s, f) in n.inputs.items())))
    super().__call__(tuples, d)
```

The method *def _uid_or_str(node_or_entity)* allows entities to be translated into nodes:

```
def _uid_or_str(node_or_entity):
    " " " Helper function to support the transition from 'Entitie's to
    Node's.
    " " "
    return (node_or_entity.uid if hasattr(node_or_entity, ''uid'')
            else str(node_or_entity))
```

The last line of the code is always present for the class *EnergySystem*.

```
DEFAULT = Grouping(_uid_or_str)
" " " The default :class:'Grouping'.

This one is always present in a :class:'energy system
<oemof.core.energy_system.EnergySystem>'. It stores every :class:'entity
<oemof.core.network.Entity>' under its :attr:'uid
<oemof.core.network.Entity.uid>' and raises an error if another :
class:'entity
<oemof.core.network.Entity>' with the same :attr:'uid
<oemof.core.network.Entity.uid>' get's added to the :class:'energy system
<oemof.core.energy_system.EnergySystem>'.
'''''
```

4.1.4   *Oemof-Network*

In order to create an optimization approach for an energy supply model in OEMOF, first all entities must be created (components – sink, source, transformer - and bus) and connected to each other. A detailed description of this can be found in Sect. 2.3, Fig. 2.7 where as an example the energy supply model for Rostock was presented. Here below you can see the code for this example (oemof-Team 2017a):

```python
from oemof.network import Entity
from oemof.energy_system import bus, sink, source, transformer, storage

# create the energy system
# The following line creates an instance 'es' of the class 'EnergySystem'.

es = EnergySystem()

# First the buses are created. These connect the components to each other.

# create bus 1
bus_1 = Bus(label="r1_gas")

# create bus 2
bus_2 = Bus(label="r1_el")

# create bus 3
bus_3 = Bus(label="r1_th")

# create bus 4
bus_4 = Bus(label="r1_bio")

# create bus 5
bus_5 = Bus(label="r2_el")

# create bus 6
bus_6 = Bus(label="r2_gas")

# create bus 7
bus_7 = Bus(label="r2_coal")

# create bus 8
bus_8 = Bus(label="r2_th")
```

```
# Next the sinks are defined.
# Then the sinks are connected to buses. This defines the inputs and
# outputs.

# create sink 1
# Empty square brackets "[]" create an empty list.
Sink(label='de1', inputs={bus_1: []})

# create sink 2
Sink(label='dh1', inputs={bus_3: []})

# create sink 3
Sink(label='de2', inputs={bus_5: []})

# create sink 4
Sink(label='dh2', inputs={bus_8: []})

# Next the sources are defined. These have precisely one output.
# This output is connected to a bus.

# create source
Source(label='wt1', outputs={bus_2: []})

# Next transformers are connected to buses and their inputs and outputs
# are defined.

# create transformer 1
Transformer(label='gt1', inputs={bus_1: []}, outputs={bus_2: []})

# create transformer 2
Transformer(label='cb1', inputs={bus_2: []}, outputs={bus_5: []})

# create transformer 3
Transformer(label='bg1', inputs={bus_4: []}, outputs={bus_2: []},
outputs={bus_3: []})

# create transformer 4
Transformer(label='cb2', inputs={bus_5: []}, outputs={bus_2: []})

# create transformer 5
Transformer(label='ptg2', inputs={bus_5: []}, outputs={bus_6: []})

# create transformer 6
Transformer(label='cp2', inputs={bus_7: []}, outputs={bus_5: []})
```

```
# create transformer 7
Transformer(label='chp2', inputs={bus_6: []}, outputs={bus_5: []},
outputs={bus_8 []})


# The energy storage must also be defined.
# create storage
Storage(label='sp1', inputs={bus_2: []}, outputs={bus_2: []})
```

We have already described how grouping is used in OEMOF. Often a grouping combines instances of one class or a specific type of a class, for example all flows with the attribute *investment*. To be able to do this, a function is required which computes a key for each group and each element of a group from the class *Entity* (*class:'entity <oemof.core.network.Entity>*). Then the group can be stored together with the key in the class *entity* under the grouping criterion type (*class:'entity <oemof.core.network.Entity>*). This can be seen in the following example:

```
# In the first part of the code a grouping according to type 'Sink' is
# carried out. The term 'range()' describes a function in Python, which
# creates lists which are arithmetic in nature and thus correspond to
# arithmetic enumerations [Klein o.J.-c]. This can be used for
# loops. In this example 'for i in range(9)' nine numbers starting with the
# number '0' are enumerated.
# The term 'format' represents the string method 'format'. This method
# generates a string, which lists the buses that have been defined.
# As a representative selection, the example from
# [Klein o.J.-b] is shown here:

>>> "First argument: {0}, second: {1}".format(47,11)
'First argument: 47, second: 11'

    es = EnergySystem(groupings=[type])
    buses = set(Bus(label="Bus {}".format(i)) for i in range(9))
    es.add(*buses)
    components = set(Sink(label=''Component {}''.format(i))
                    for i in range(9))
    es.add(*components)
    buses == es.groups[Bus]
True
    components == es.groups[Sink]
True
```

Groupings are created automatically in OEMOF. We have seen this time and again in the source code excerpts above.

Other methods in this module are required for the smooth running of the program, but are not relevant to the modeling itself, which is why they are not discussed in detail here.

Once the energy system with all its nodes, components, flows etc. has been created, the further description of the model is carried out by the oemof.solph library.

4.1.5 Oemof-Solph

The oemof.solph library provides tools and components for creating linear optimization models. These can be applied both to purely linear models and to mixed-integer linear optimization models. This library also allows modelers to switch between dispatch models, i.e., models that are concerned with finding an optimal operation strategy for power stations and investment models looking for lowest possible investment costs. Creating a model that combines both aspects is also possible. By means of an investment cost model, users can determine if maintaining an existing power station is better value for money than replacing it by a newer model. For such a model, an annuity cost calculation will be used. In some cases, minimization of costs may not be the only relevant parameter, other aspects, such as the amount of emissions might also be relevant. oemof.solph uses the Python package Pyomo for formulating an optimization problem.

To start creating a model, first a datetime index must be set via the class *pandasDatetimeIndex* in order to create an energy system (pandas o.J.-a). The class *pandasDatetimeIndex* is used to describe the overall time period analyzed as well as the separate timesteps. Pandas provides the function *date_range* for this. Further information on this function can be found in (pandas 2017b). This lists parameters for the DatetimeIndex. There are three key parameters: *start*, *end* and *period*, of which at least two must be defined. The data types for these are shown in Table 4.2.

Table 4.2 Additional information on the parameters for the pandas.date_range (pandas 2017b)

Parameter	Data type/value	Additional information
Start	string or datetime-like, default None	Left bound for generating dates.
End	string or datetime-like, default None	Right bound for generating dates.
Periods	integer, default None	Number of periods to generate.
Freq	string or DateOffset, default 'D' (calendar daily)	Frequency strings can have multiples, e.g., '5H'.
tz	string, default None	Time zone name for returning localized DatetimeIndex, for example Asia/Hong_Kong.
Normalize	bool, default False	Normalize start/end dates to midnight before generating date range.
Name	string, default None	Name of the resulting DatetimeIndex.
Closed	string, default None	Make the interval closed with respect to the given frequency to the 'left', 'right', or both sides (None).

The parameter *freq* has several aliases for its data type so-called 'offset aliases', which are listed in (pandas 2017c). For example *H* represents an hourly frequency and *S* a second-by-second frequency.

To set the datetime index, the following information must be entered into the source code for the year 2017 in our example below, using hourly timesteps.

```
# First, Pandas must be imported:
import pandas as pd
# Next the datetime index must be set. The start point for this is
# 1/1/2017 and the period, here the overall number of hours for the year
# is given as 8760 h.
# In addition to this, the frequency is listed in hours (H).
my_index = pd.date_range('1/1/2017', periods=8760, freq='H')
```

The output returned is based on the following code:

```
return DatetimeIndex(start=start, end=end, periods=periods,
                     freq=freq, tz=tz, normalize=normalize, name=name,
                     closed=closed, **kwargs)
```

Using this datetime index the energy system can now be created.

```
import oemof.solph as solph
my_energysystem = solph.EnergySystem(timeindex=my_index)
```

Once this step has been completed, the components can be created.

Defining buses in solph

In OEMOF buses have inputs and outputs, which are represented in the model by flows. The inputs and outputs of a bus must always be balanced. In order to define an instance of a bus, first a unique name must be given by means of a *label*. One instance of our Rostock example has the unique name *r1_bio*. The following excerpt shows how to define an instance of a bus (oemof-Team 2017a):

```
solph.Bus(label='r1_bio')
```

It would equally be possible to use a variable, such as *electricity*, which could be assigned to the object. This could facilitate connecting it to a specific component (oemof-Team 2017a):

```
electricity_bus = solph.Bus(label='electricity')
```

The modelers can then query the output of the instance *r1_bio*. This example shows what difference including a variable makes for the programming code. (oemof-Team 2017a):

```
print(my_energsystem.groups['r1_bio']
print(electricity_bus)
```

The fact that a variable has been defined can sometimes reduce the amount of source code required.

The class flow

The class *Flow* connects entities with each other and also represents the edges in a graph model. By means of the class *Flow* investments and other cost flows but also emissions can be represented as flows. These flows are normally connected to components and are therefore defined together with the components. Flows have upper and lower limits (bounds), which can be either time-dependent or time-independent (constant). Sometimes combined limits can also be set. Defining a flow is very easy in OEMOF (oemof-Team 2017b):

```
solph.Flow()
```

Defining a sink in OEMOF

Energy demand within an energy supply system is modeled by means of a sink. But a sink can also be used to take excess energy out of the system.

Following on from the definition of a bus with its variable *electricity_bus* the following section describes the definition of a sink in oemof.solph. A sink is an energy consumer, such as a multi-family house or a manufacturing site. In our Rostock example, a specific instance of a sink might be *de1*. As the bus connected to this sink must still be balanced, the code example gives the information, that object *de1* receives input from the bus introduced earlier with its variable *electricity_bus* (oemof-Team 2017b):

```
solph.Sink(label='de1', inputs={electricity_bus: solph.Flow(
              actual_value=my_demand_series, fixed=True,
              nominal_value=nominal_demand)})
```

The parameter *my_demand_series* contains a sequence of normalized values for demand. Thus the parameter *nominal_value* is created from the product of the maximum demand with the relevant sequence of normalized values. The parameter *fixed=True* ensures that the *actual_value* cannot be changed by the solver.

An excess sink is less restricted, as this can be used to take on all excess production within a system (oemof-Team 2017b):

```
solph.Sink(label='electricity_excess', inputs={electricity_bus:
            solph.Flow()})
```

The *Sink* class takes on all energy provided in a system. Thus sinks could be seen as one giant plug open for consumption. There are no further restrictions and no more variables.

Defining a source in oemoph-solph

In OEMOF a source could be a wind power plant or a solar power plant. But an import of gas or oil could also be a source. It is also possible to define a so-called slack variable via this class, in order to balance energy flows within a system and to ensure that the optimization problem can be solved.

In our example Rostock, the sources are a wind power plant and a gas turbine. For the wind power plant weather data can be defined per hour as feed-in. For the gas turbine the fuel gas can be imported, which is limited by certain restrictions. Thus for the gas import a maximum can be defined (maximum value) by means of the parameter *nominal_value*. This defines the amount of gas that can be imported per hour. But in the case of a wind power plant the parameter *nominal_value* could also indicate the installed output maximum. Another type of restriction possible is an annual limit. This can be expressed by the parameter *summed_max*.

Importing gas to a system produces costs. These are entered into the system via the parameter *variable_cost* in oemof.solph (oemof-Team 2017b). The costs are given in hourly timesteps:

```
solph.Source(
    label='import_natural_gas',
    outputs={my_energsystem.groups['r1_gas']: solph.Flow(
        nominal_value=1000, summed_max=1000000, variable_costs=50)})
solph.Source(label='wt1', outputs={r1_el: solph.Flow(
        actual_value=wind_power_feedin_series, nominal_value=1000000,
        fixed=True)})
```

A source is defined very similarly to a sink. The main difference is that sources add energy to a system while sinks take it out of a system. But sources also do not have any more variables or restrictions.

The class transformer in oemof-solph

As OEMOF works by and large as a linear optimization approach, transformer processes must be viewed as constant in terms of time. This means that the efficiency of a source etc. must be defined as constant across the defined period. But it is possible to define different degrees of efficiency for different periods. This could be the case for a CHP plant, which can have different levels of efficiency across different operation settings (full load, half load).

This class is called *Transformer* in OEMOF. The class *Transformer* has been reworked in the new Release 0.2.0 and is therefore no longer compatible with the previous Releases 0.1.x. The definitions and descriptions of Sect. 3.1.4 all refer to the new release. But as some important studies have been published based on the old *Transformer* class from older OEMOF versions, some information on the previous versions is given here, so that you can follow the previous studies and concepts, such as (Gaudschau 2017).

According to the Release Versions 0.1.x a transformer can be one of two different components. The first type has one input and one or more outputs. One such type of transformers could be a CHP plant. The second type has one or more inputs and one output. This would be the case for a heat pump. For this reason the old Release Versions distinguish between Transformer type 1 (1xM) and Transformer type 2 (Nx1). The term in brackets shows that type 1 (1xM) has only one input and *m* outputs, while type 2 (Nx1) has *n* inputs and one output.

The new Release 0.2.0 has only one type, the class *Transformer*. This class is designed to take on both types, as components can be described that have *n* inputs and *m* outputs. This release thus allows modelers to define transformers that have several inputs and several outputs.

In principle, a transformer can represent any type of transformation within an energy system, such as a power plant, an electricity cable, an electrolysis plant or a cooling device.

Let us first look at a *Transformer* (1xM) as described in Release 0.1.x. In the Rostock example, a gas turbine is described in the following way according to Release 0.1.x:

```
solph.LinearTransformer(
    label='gt1',
    inputs={my_energsystem.groups['r1_gas']: solph.Flow()},
    outputs={r1_el: solph.Flow(nominal_value=10e10)},
    conversion_factors={r1_el: 0.58})
```

A CHP (combined heat and power plant, which was already mentioned in the context of the biogas example) would be described in a similar manner in Release 0.1.x:

```
# First the two buses for the outputs are introduced.
r1_el = solph.Bus(label='electricity')
r1_th = solph.Bus(label='heat')
# Next the LinearTransformer(1xM) is defined.
solph.LinearTransformer(
    label='bg1',
    inputs={r1_bio: Flow()},
    outputs={r1_el: Flow(nominal_value=30),
             r1_th: Flow(nominal_value=40)},
             conversion_factors={r1_el: 0.3, r1_th: 0.4})
```

What about the second option, a transformer with more than one input (Nx1)? Again the efficiency rate, or, in the case of a heat pump the Coefficient of Performance (COP), must be constant within one timestep. However, for each separate timestep a different efficiency rate can be set. But this setting must be determined before the optimization model is computed and cannot be changed during the computation.

Let us have a look at a heat pump as an example. As heat pumps have not been described in our Rostock example, we must first introduce new buses for the heat pump. The inputs for a heat pump are low temperature heat and electricity, the output is high temperature heat. A new object of a transformer (Nx1) is created. This object is given the label *heat_pump* in the model (Release 0.1.x):

```
# First the three buses are created (input: electricity, low temperature
# heat, output: high temperature heat.
b_el = solph.Bus(label='electricity')
b_th_low = solph.Bus(label='low_temp_heat')
b_th_high = solph.Bus(label='high_temp_heat')

# Next the Coefficient of Performance of the heat pump is defined.
cop = 3

# Next this transformer is defined as a LinearN1Transformer.
# Next the object heat pump is initialized.
solph.LinearN1Transformer(
    label='heat_pump',
    inputs={bus_elec: Flow(), bus_low_temp_heat: Flow()},
    outputs={bus_th_high: Flow()},
            conversion_factors={bus_elec: cop, b_th_low: cop/(cop-1)})
```

The other type of transformers is those that have one input, but two outputs. The particularity of these is the variable relation between the two outputs. If we look at a CHP plant, the two outputs are heat and electricity. There are two possible operation modes for a CHP plant. If both outputs are required equally, then the relation between the two outputs is fixed before the plant starts operation. This would mean that the proportion of heat and electricity are fixed from the outset. For this mode of operation, back-pressure turbines are ideal. But if the demand fluctuates, for example, if heat and electricity are required during the week, but only electricity at the weekend, this means that the plant is used in variable mode, the proportion of heat and electricity produced varies. For this variable operation mode extraction-condensation turbines are often used. But the variable and fixed operation modes have different COPs, which must be entered correctly into the calculations.

In order to reflect this in the model, the class *VariableFractionTransformer* can be used. By modeling one main flow and one separate extraction flow (tapped flow) a variable distribution between the two output flows can be programmed. In our case the main output flow is electricity, as this is produced throughout the week, and the tapped flow is the heat extraction, as this is not required at the weekend. Now the efficiency rate for the main flow can be set as a separate parameter, e.g., as *conversion_factor_single_flow*.

In our example Rostock the first of the biogas CHP plants *r1_el* can be modeled as the main flow and the second bus *r1_th* as secondary flow. To indicate that the flow of electricity and heat vary, the object biogas CHP is given the extra label *variable (variable_bg1)* (oemof-Team 2017b).

```
solph.VariableFractionTransformer(
    label='variable_bg1',
    inputs={r1_bio: solph.Flow(nominal_value=10e10)},
    outputs={r1_1: solph.Flow(), r1_th: solph.Flow()},
            conversion_factors={r1_el: 0.3, r1_th: 0.5},
            conversion_factor_single_flow={r1_el: 0.5}
            )
```

In the new Release 0.2.0 the class *Transformer* can have *n* inputs and *m* outputs. Thus the distinction between transformaters of type 1 (Nx1) and transformers of type 2 (1xM) is no longer needed.

Additional class storage

Storage units are not part of the classes defined in the module *oemof.network*. This class exists only in *oemof.solph*.

Storage units have some specific characteristics. A storage unit has a capacity, which is defined via the parameter *nominal_value*. As storage units have limited input and output flows depending on their capacity, it is relevant to set a ratio for flow and capacity. For this reason the parameter *nominal_input_capacity_ratio*, can be set for the input and a parameter like *nominal_output_capacity_ratio* can be created. Further parameters relevant for storage units are their efficiency rate during charging and discharging as well as the loss per timestep (conversion factor). The storage unit in our Rostock example is pasted into the source code as follows:

```
# Storage is created in solph. As an additional flow variable
# costs for input and output flows are defined.
solph.Storage(
    label='storage',
    inputs={r1_el: solph.Flow(variable_costs=10)},
    outputs={r1_el: solph.Flow(variable_costs=10)},
            capacity_loss=0.001, nominal_value=50,
            nominal_input_capacity_ratio=1/6,
            nominal_output_capacity_ratio=1/6,
            inflow_conversion_factor=0.98, outflow_conversion_factor=0.8)
```

Setting up an optimization model in solph

As described earlier, OEMOF can be used either to create models for the optimization of costs or for the optimization of an operation strategy. But costs could also be formulated in emissions rather than in investment costs. Investment cost calculations are described in more detail in Sect. 6.1.

In order to set up an optimization model in *Solph*, the modeler must first enter what type of optimization is to be modeled. If a dispatch model is to be created, the first step is to import the library *os*. Then the information that a dispatch model is to be created must be entered into *Solph*.

```
import os
# There is a simple least cost optimization model (OperationalModel).
om = solph.OperationalModel(my_energysystem)

# By means of the solver CBC costs are to be minimized.
om.solve(solver='cbc', solve_kwargs={'tee': True})
```

Within your model an instance (object) of the class *Investment* is created. This instance can be added to the components analyzed in the model. Investment can also be added to storages or flows. If parameters have been defined within the model that refers to the parameter *nominal_value* or *nominal_capacity* these now report to the variable *investment*. For this reason the object investment connected to a component should not have the parameter *nominal_value* or *nominal_capacity*.

In our Rostock example the object *investment* could be used to analyze the optimal capacity of wind turbine *wt1* to minimize overall costs to the energy system.

```
solph.Source(label='wt1', outputs={r1_el: solph.Flow(
             actual_value=wind_power_time_series, fixed=True,
             investment=solph.Investment(ep_costs=epc, maximum=100000))})
```

Several parameters are defined and used for periodic costs. The parameter *capex* (capixpenditure) is used to define the investment for a long-term asset, such as a plant, a machine, or a building, which are normally financed by a mixture of debt capital and equity capital. The parameter *wacc* (Weighted Average Cost of Capital) is composed of the arithmetic mean of the debt and equity capital costs of a company or, as is the case in our example, an investment. Of course, the weighting must correspond to the actual weight of debt capital and equity capital (Gabler 2018). The weighting for our example might look like this:

```
# The investment cost
capex = 2000
# The life expectancy
```

```
lifetime = 25
# The weighted average capital cost (wacc)
wacc = 0.05
# Calculation of epc.
epc = capex * (wacc * (1 + wacc) ** lifetime) / ((1 + wacc) **
      lifetime - 1)
```

Let us try to define investment costs for a storage unit in our Rostock example:

```
solph.Storage(
    label='st1', capacity_loss=0.01,
    inputs={r1_el: solph.Flow()},
    outputs={r1_el: solph.Flow()},
            nominal_input_capacity_ratio=1/6,
            nominal_output_capacity_ratio=1/6,
            inflow_conversion_factor=0.99,
            outflow_conversion_factor=0.8,
    investment=solph.Investment(ep_costs=epc))
```

Selection of a mixed-integer linear problems

In order to model a MIP OEMOF provides the *oemof.solph.options module*. This module contains the two MIP classes *BinaryFlow* and *DiscreteFlow*. But in Release 0.2.0 these are not yet compatible with the *Investment* class and can therefore not be combined yet. In order to use these two classes, these are called as part of the Flow declaration.

```
r1_el = solph.Bus(label='electricity')
r1_th = solph.Bus(label='heat')
solph.LinearTransformer(
    label='bg1',
    inputs={r1_bio: Flow(discrete=DiscreteFlow())},
    outputs={r1_el: Flow(nominal_value=30, binary=BinaryFlow()),
            r1_th: Flow(nominal_value=40)},
            conversion_factors={r1_el: 0.3, r1_th: 0.4})
```

This defines the input variable *Flow* of the LinearTransformer *bg1* as discrete variable. This means that this variable could have a value from (min, ..., 5, 6, 7, ..., max). For the output the variable of the bus *r1_el* is defined as binary variable. This means that value 1 indicates an electricity flow and value 0 indicates no electricity production.

The module groupings

Groupings have been described in several chapters. The main introduction is given in Sect. 4.13.

Other modeling examples:

The following examples all refer to Release 0.2.0. This means that the two separate transformer types type 1 (1xM) and type 2 (Nx1) no longer exist, instead, there is simply the class *Transformer*.

In the example, a transmission line is modeled in the invesment mode, using both types of investment variables. The equivalent periodical costs (epc) of line are '20', what are currency units here. These currency units must be defined beforehand, for example in Euros. As a tryout, this value can be assigned to one transmission line first, the others are left at zero and then you could have a look at the result (GitHub 2017b):

```python
import pandas as pd
    from oemof import solph
    date_time_index = pd.date_range('1/1/2017', periods=5, freq='H')
    energysystem = solph.EnergySystem(timeindex=date_time_index)
    bel1 = solph.Bus(label='electricity1')
    bel2 = solph.Bus(label='electricity2')
    energysystem.add(bel1, bel2)
    energysystem.add(solph.Transformer(
                label='powerline_1_2',
                inputs={bel1: solph.Flow()},
                outputs={bel2: solph.Flow(
                    investment=solph.Investment(ep_costs=20))}))
    energysystem.add(solph.Transformer(
                label='powerline_2_1',
                inputs={bel2: solph.Flow()},
                outputs={bel1:
                solph.Flow(
                    investment=solph.Investment(ep_costs=20))}))
    om = solph.Model(energysystem)
    line12 = energysystem.groups['powerline_1_2']
    line21 = energysystem.groups['powerline_2_1']
    solph.constraints.equate_variables(
                om,
                om.InvestmentFlow.invest[line12, bel2],
                om.InvestmentFlow.invest[line21, bel1])
```

The following example based on the class *Link(Transformer)* shows how a class is defined with its attributes, but also how to give a print command.

```python
from oemof import solph
    bel0 = solph.Bus(label='el0')
    bel1 = solph.Bus(label='el1')
```

```
link = solph.custom.Link(
        label='transshipment_link',
        inputs={bel0: solph.Flow(), bel1: solph.Flow()},
        outputs={bel0: solph.Flow(), bel1: solph.Flow()},
        conversion_factors={(bel0, bel1): 0.92, (bel1, bel0): 0.99})
    print(sorted([x[1][5] for x in link.conversion_factors.items()]))
[0.92, 0.99]
    type(link)
     <class 'oemof.solph .custom.Link'>
    sorted([str(i) for i in link.inputs])
['el0', 'el1']
    link.conversion_factors[(bel0, bel1)][3]
0.92
```

Next two examples using the *Flow* class show how to define a fixed flow and how to model a class with an upper and lower limit. First the fixed flow:

```
f = Flow(actual_value=[10, 4, 8], fixed=True, variable_costs=8)
```

If a flow is to have an upper or lower limit, this can be modeled in the following way. Here the maximum value of *f1* is queried:

```
f1 = Flow(min=[0.2, 0.3], max=0.99, nominal_value=100)
```

4.1.6   Oemof-Outputlib

The output of results is supported by the Pandas package, which is part of OEMOF. (pandas 2017a) gives more information as to how the Pandas toolkit can be used to analyze data.

The *Outputlib* has been updated and extended for Release v0.2.0. The new library does no longer contain tools for plotting optimization results. This part has been taken over by the new visualization package oemof_visio. This can be found on the Github homepage at: https://githqub.com/oemof. The next section will give some more detailed information.

The library *oemof-outputlib*

The library *oemof-outputlib* is mainly responsible for gathering and organizing output data in the new Release. Data are organized in the form of a Python dictionary. This contains all scalar values as Pandas series. All nodes and their flows are returned as Pandas DataFrames.

The sequences and scales model nodes. These are provided with an internal key, such as (node, None). The flows are also assigned keys, each referring to a tuple such as (node_1, node_2). Data can be read straight from the dictionary via these keys. *Node* is the name of the object which is to be addressed.

How are the data collected? This is what the processing module is for (GitHub 2018a). The module is called via the following programming code (OEMOF 2014):

```
results = outputlib.processing.results(om)
```

The result are the data. If you just wanted to find out which flows have been created, you could ask for the keys. This is shown in the following code excerpt:

```
energysystem.results['main'] =
outputlib.processing.results(optimization_model)
energysystem.results['meta'] =
outputlib.processing.meta_results(optimization_model)
print(energysystem.results['main'].keys())
dict_keys([(<oemof.solph .network.
Transformer object at 0x7f6b44195138>, <oemof.solph .network.
Bus object at 0x7f6b44125188>), (<oemof.solph .network.
Transformer object at 0x7f6b4414ccc8>, <oemof.solph .network.
Bus object at 0x7f6b44125138>), (<oemof.solph .network.
Bus object at 0x7f6b44125138>, <oemof.solph .network.
Sink object at 0x7f6b4414c688>), (<oemof.solph .network.
Bus object at 0x7f6b44125188>, <oemof.solph .network.
Sink object at 0x7f6b4414c908>), (<oemof.solph .network.
Bus object at 0x7f6b4418ad68>, <oemof.solph .network.
Transformer object at 0x7f6b440dd4f8>), (<oemof.solph .network.
Bus object at 0x7f6b441250e8>, <oemof.solph .network.
Transformer object at 0x7f6b4414ccc8>), (<oemof.solph .network.
Transformer object at 0x7f6b4414cef8>, <oemof.solph .network.
Bus object at 0x7f6b44125138>), (<oemof.solph .network.
Bus object at 0x7f6b44125138>, <oemof.solph .network.
Sink object at 0x7f6b44138ea8>), (<oemof.solph .network.
Source object at 0x7f6b441484f8>,
# ... more data follows, the list was stopped here.
```

As the keys are hard to read in code, they can be converted to strings. The conversion of keys into readable strings can be seen in this code excerpt:

```
string_results =
outputlib.views.convert_keys_to_strings(energysystem.results['main'])
print(string_results.keys())
dict_keys([('pp_chp', 'bth'), ('pp_lig', 'bel'), ('bel', 'heat_pump'),
```

```
('bel', 'demand_el'), ('bth', 'demand_th'), ('oil', 'pp_oil'),
('lignite', 'pp_lig'), ('pp_coal', 'bel'), ('bel', 'excess_el'),
('pv', 'bel'), ('gas', 'pp_chp'), ('pp_gas', 'bel'), ('gas', 'pp_gas'),
('pp_oil', 'bel'), ('heat_pump', 'bth'), ('wind', 'bel'),
('b_heat_source', 'heat_pump'), ('heat_source', 'b_heat_source'),
('pp_chp', 'bel'), ('coal', 'pp_coal')])
```

The next step is to export the data in form of a table. This code excerpt gives the necessary command for this:

```
df = node_results_bel['sequences']
df.head(2)
```

The term *df* stands for 'Pandas DataFrame'. At this point the Pandas DataFrame table of results *node_results_bel['sequences']* is passed to *df*.

The command *df.head(2)* returns the first two rows of the DataFrame as shown in Table 4.3.

It is also possible to create your own results table. This is the relevant command:

```
d = {'high' : pd.Series([3., 2., 1.], index=['a', 'b', 'c']),
     'low' : pd.Series([4., 3., 2., 1.], index=['a', 'b', 'c', 'd'])}

df = pd.DataFrame(d)
```

The term *d* is the name of the object which contains the output data. The output data look like this:

```
   high   low
a   3.0    4.0
b   2.0    3.0
c   1.0    2.0
d   NaN    1.0
```

As mentioned before, OEMOF collects data as groupings. This can be used for the return of output data by giving the command via *Groupings*. For this the name of the grouping needs to be called, such as the group *wind*:

```
node_wind = energysystem.groups['wind']
print(results[(node_wind, bus_electricity)])
```

Table 4.3 Table of Results returned via outputlib

	((bel, demand_el), flow)	((bel, excess_el), flow)	((bel, heat_pump), flow)	((pp_chp, bel), flow)	((pp_coal, bel), flow)	((pp_gas, bel), flow)	((pp_lig, bel), flow)	((pp_oil, bel), flow)	((pv, bel), flow)	((wind, bel), flow)
2017–01–01 00:00:00	52.169653	0.0	0.821571	2.559823	20.2	0.0	11.8	0.0	0.0	18.4314
2017–01–01 01:00:00	52.169653	0.0	0.911379	2.450731	20.2	0.0	11.8	0.0	0.0	18.6303

Toolkit oemof_visio

In addition to outputlib another key element of the output display is the new toolkit *oemof_visio* for visualizing the results. The first step is installing the *oemof_visio* toolkit. This is done by using *pypi* using the command *pip install*:

```
pip install git+https://github.com/oemof/oemof_visio.git
```

Further information can be found at: https://github.com/oemof/oemof_visio. If you receive an error message, this means that the git program has not been installed yet:

```
Error [WinError 2] ... while executing command git clone -q https://github.
com/oemof/oemof_visio.git C:\Users\...\AppData\Local\Temp
\pip-rkdz0ake-build Cannot find command 'git'
```

This can still be done at this point. One way of installing can be found at https://git-scm.com/.

The directory *oemof_visio* contains the package *plot.py* which provides tools for the output of data such as:

- Slicing the DataFrame
- Plotting parts of the DataFrame
- Creating a color dictionary.

See also (oemof 2018a).

There are a few examples for doing this in Release v0.2.0 (at: https://github.com/oemof/oemof_examples/tree/master/examples/oemof_0.2) you can find a directory with plotting options, cf.: https://github.com/oemof/oemof_examples/tree/master/examples/oemof_0.2/plotting_examples. As an example of presenting plotting results the example *storage_investment_plot.py* was chosen (cf.: https://github.com/oemof/oemof_examples/blob/master/examples/oemof_0.2/plotting_examples/storage_investment_plot.py). Once the models have been implemented they are called applications.

The command requesting plotting in oemof_visio is *view*:

```
# Getting results and views
results = processing.results(om)
custom_storage = views.node(results, 'storage')
electricity_bus = views.node(results, 'electricity')
```

The application *storage_investment_plot.py* returns data for *custom_storage* and for the *electricity_bus*. The following section will introduce several options for plotting data.

The first option is to plot the balance for the component *custom_storage*. The results in graph (a) are shown as colored lines, as indicated in Fig. 4.1. Graph (b) changes the datetime ticks.

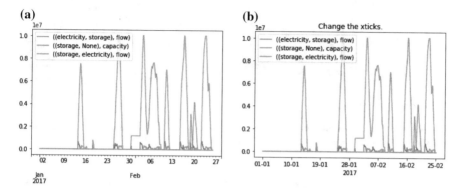

Fig. 4.1 Option 1: Plotting of component *custom_storage* as line. **a)** Plot directly using pandas. **b)** Change the datetime ticks (oemof 2018b)

In order to generate the graphs shown above, the following code input is required (oemof 2018b). First plot option (a):

```
# ***** 1. Example a) ***************************************************
# Plot directly using pandas
custom_storage['sequences'].plot(kind='line', drawstyle='steps-post')

plt.show()
```

Now for plot option (b):

```
# ***** 1. Example b) ***************************************************
# Change the datetime ticks
ax = custom_storage['sequences'].reset_index(drop=True).plot(
    kind='line', drawstyle='steps-post')
ax.set_xlabel('2017')
ax.set_title('Change the xticks.')
oev.plot.set_datetime_ticks(ax, custom_storage['sequences'].index,
                    date_format='%d-%m', number_autoticks=6)

plt.show()
```

In the next graph Fig. 4.2, which corresponds to the second plotting option, the balance is for *electricity_bus*. This graph uses different colors.

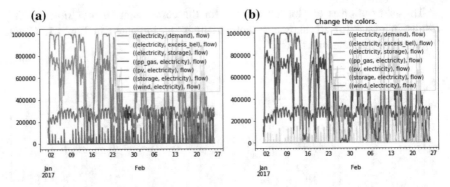

Fig. 4.2 Presentation of the results for *electricity_bus* as colored lines. **a)** Plot directly using pandas. **b)** with colors changed according to dictionary (oemof 2018b)

For the graph in Fig. 4.2 find the programming source code below, the first step is to return all flows and components of the model:

```
# ***** 2. Example a) *********************************************
cdict = {
    (('electricity', 'demand'), 'flow'): '#ce4aff',
    (('electricity', 'excess_bel'), 'flow'): '#555555',
    (('electricity', 'storage'), 'flow'): '#42c77a',
    (('pp_gas', 'electricity'), 'flow'): '#636f6b',
    (('pv', 'electricity'), 'flow'): '#ffde32',
    (('storage', 'electricity'), 'flow'): '#42c77a',
    (('wind', 'electricity'), 'flow'): '#5b5bae'}


# Plot directly using pandas
electricity_bus['sequences'].plot(kind='line', drawstyle='steps-post')

plt.show()
```

The next step is to change the colors:

```
# ***** 2. Example b) *********************************************
cdict = {
    (('electricity', 'demand'), 'flow'): '#ce4aff',
    (('electricity', 'excess_bel'), 'flow'): '#555555',
    (('electricity', 'storage'), 'flow'): '#42c77a',
    (('pp_gas', 'electricity'), 'flow'): '#636f6b',
    (('pv', 'electricity'), 'flow'): '#ffde32',
    (('storage', 'electricity'), 'flow'): '#42c77a',
    (('wind', 'electricity'), 'flow'): '#5b5bae'}
```

```
# Change the colors using the dictionary above to define the colors
colors = oev.plot.color_from_dict(cdict, electricity_bus['sequences'])
ax = electricity_bus['sequences'].plot(kind='line', drawstyle='steps-post',
color=colors)
ax.set_title('Change the colors.')

plt.show()
```

The third plotting option again analyzes the balance for *electricity_bus*. This time the input flows are returned (cf. Fig. 4.3).

Fig. 4.3 Plotting results for *electricity_bus* as lines, selecting only input flows (oemof 2018b)

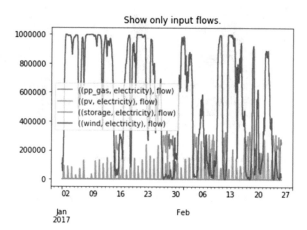

The following source code is used to create the graph:

```
# ***** 3. example *********************************************
# Plot only input flows
in_cols = oev.plot.divide_bus_columns(
    'electricity', electricity_bus['sequences'].columns)['in_cols']
ax = electricity_bus['sequences'][in_cols].plot(kind='line',
                                       drawstyle='steps-post')
ax.set_title('Show only input flows.')

plt.show()
```

The term *oev* represents the oemof_visio package. This is used to call up the plotting tool from oemof_visio. The next term *divide_bus_columns* creates a dictionary with two separate lists. One list contains all inputs, the other contains all outputs. These are passed to the dataframe from which the relevant list can be returned, in this case the list of input flows. To receive this the name of the list of all input flows, *in_cols*, must be entered.

The third plotting option is to plot the data in form of stacked plains (cf. Figure 4.4). The data plotted represent again the balance for *electricity_bus*.

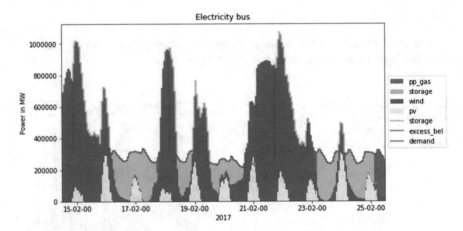

Fig. 4.4 Presentation of results for *electricity_bus* as stacked plains (oemof 2018b)

The source code requires the following changes:

```
# ***** 4. example **************************************************
# Create a plot to show the balance around a bus.
# Order and colors are customisable.

inorder = [(('pv', 'electricity'), 'flow'),
           (('wind', 'electricity'), 'flow'),
           (('storage', 'electricity'), 'flow'),
           (('pp_gas', 'electricity'), 'flow')]

fig = plt.figure(figsize=(10, 5))
electricity_seq = views.node(results, 'electricity')['sequences']
plot_slice = oev.plot.slice_df(electricity_seq,
                               date_from=pd.datetime(2017, 2, 15))
my_plot = oev.plot.io_plot('electricity', plot_slice, cdict=cdict,
                           inorder=inorder, ax=fig.add_subplot(1, 1, 1),
                           smooth=False)
ax = shape_legend('electricity', **my_plot)
oev.plot.set_datetime_ticks(ax, plot_slice.index, tick_distance=48,
                            date_format='%d-%m-%H', offset=12)
```

```
ax.set_ylabel('Power in MW')
ax.set_xlabel('2017')
ax.set_title("Electricity bus")
```

The command *inorder* dictates the order of the plains in the stack. The flow of component *pv* is at the bottom followed by the flows of the components *wind*, *storaga* und *pp_gas*. In addition to this the y axis is given the label *Power in MW*.

A simplified graph presents the stacked plains as smooth. The result can be seen in Fig. 4.5.

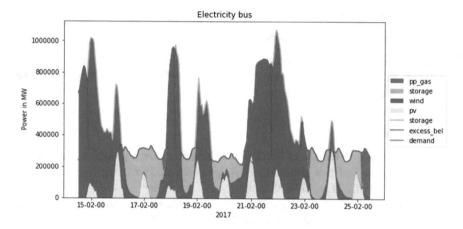

Fig. 4.5 Presentation of *electricity_bus* as stacked plains with smooth borders (oemof 2018b)

The source code required to produce Fig. 4.5 is given below:

```
# ***** 5. example *************************************************
# Create a plot to show the balance around a bus.
# Make a smooth plot even though it is not scientifically correct.

inorder = [(('pv', 'electricity'), 'flow'),
           (('wind', 'electricity'), 'flow'),
           (('storage', 'electricity'), 'flow'),
           (('pp_gas', 'electricity'), 'flow')]

fig = plt.figure(figsize=(10, 5))
electricity_seq = views.node(results, 'electricity')['sequences']
plot_slice = oev.plot.slice_df(electricity_seq,
                               date_from=pd.datetime(2017, 2, 15))
my_plot = oev.plot.io_plot('electricity', plot_slice, cdict=cdict,
                           inorder=inorder, ax=fig.add_subplot(1, 1, 1),
                           smooth=True)
```

```
ax = shape_legend('electricity', **my_plot)
ax = oev.plot.set_datetime_ticks(ax, plot_slice.index, tick_distance=48,
                                 date_format='%d-%m-%H', offset=12)

ax.set_ylabel('Power in MW')
ax.set_xlabel('2017')
ax.set_title("Electricity bus")

plt.show()
```

The attribute *smooth* must be set to *True*.

Representation of the graph structure

There is another very useful program. It is possible to have the graph structure (network structure) of the model plotted in OEMOF. Two packages must be installed to allow you to plot the network plan:

– NetworkX
– PyGraphviz.

NetworkX is a Python package. It is also an open source program. This can be used to generate and plot complex network structures. But it can also be used to analyze and modify network structures.

The package NetworkX is installed via the command *pip install*. The program system NetworkX can be downloaded at: https://pypi.python.org/pypi/networkx/2.1.

The program PyGraphviz is part of the program system Graphviz, which is also an open source program (Graphviz o.J.). This was developed to visualize directed and non-directed graphs. The program uses the labeling langauge DOT.

PyGraphviz was developed specifically for Python programs. It works as an interface for Graphviz (PyGraphviz developer team 2016). By means of this program graphs can be created, modified and plotted in Python.

PyGraphviz can run under *Windows* and under several other operating systems. But installing PyGraphviz is not that simple, as it cannot be directly installed in Windows. This chapter cannot give a specific recommendation on how to run this program in Windows. One option is to first install Graphviz. This can be installed for Windows systems via this link: https://graphviz.gitlab.io/download/. Following the *Windows* tab you will get to several links, such as *Stable 2.38 Windows* install packages which provide installation options. Version *graphviz-2.38.msi* works very well for the installation (as of April 2018). The program suggests a directory for installation. Once the program has been installed, the path for the system variables must be adapted. Section 5.1 describes how to do this for installing OEMOF. The path must be saved to *dot.exe* which can be found in the *bin* directory of Graphviz. Then Graphviz must be installed via the WinPython Command Prompt Console. This is done by entering the command *pip install graphviz* in the WinPython Command Prompt Console. Once the program has been installed it can be used for visualizing and analyzing network structures even outside Python.

Next PyGraphviz must be installed. It can be downloaded from this link: https://
pypi.python.org/pypi/pygraphviz. If there is an error message when you try to use
it, you can add missing libraries via the program system Anaconda. This program
can be downloaded and installed from the OEMOF website: http://oemof.
readthedocs.io/en/latest/installation_and_setup.html.

Fig. 4.6 shows the plotting of an optimization problem as a network plan using
PyGraphviz.

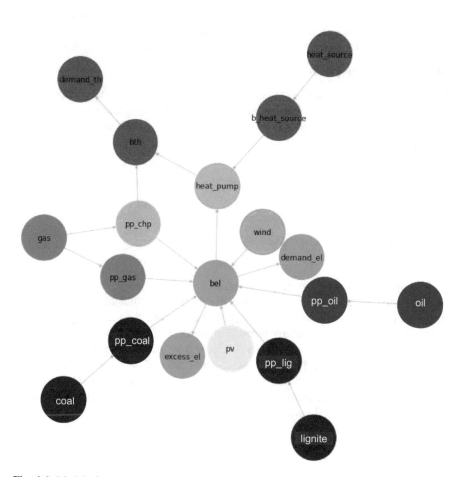

Fig. 4.6 Model of an energy system as network plan, i.e., as graph structure

4.1.7 Oemof-Feedinlib

Once the energy model has been created in OEMOF, the model must be fed with data. While the library demandlib provides data on energy demand, the library *feedinlib* provides data on the components, such as solar plants and wind power plants. Sometimes the modeler might already have data on solar and wind power plants in the form of time series. But in some cases the modelers might only have the weather data for the relevant time periods, which means that the yield per time unit must still be computed. These computations are done in the *feedinlib*. This library is part of the overall OEMOF system and can also be used as a stand-alone service. Section 4.2 describes in detail how to download the relevant data.

More on modeling yields

Several parameters can be used to compute performance data. It is very important to ensure that units are used consistently throughout the computations, in order to ensure that the computed result is accurate. The following units can be used for the listed parameters:

- Pressure (Pa)
- wind speed (m/s)
- irradiation (W/m^2)
- peak power (W)
- installed capacity (W)
- nominal power (W)
- area (m^2).

It is up to the modeler to define the units.

Each solar plant or wind power plant of any manufacturer has its own unique technical data. When modeling an energy supply system, the plants are positioned at locations that offer particularly promising yields. But this location bias can also simply be defined and modeled accordingly, it does not have to reflect an actual plant with an actual yield. But of course, the relevance of such a theoretical model may not be as strong as one that models reality. The technical data of the power plants combined with the location are key features based on which the yield can be computed. Below several parameters are listed that are required for calculating the yield of wind and solar power plants Table 4.4 lists the parameters for wind power plants.

Table 4.4 List of parameters with examples for wind power yield (oemof developing group 2015)

Parameter	Description
h_hub	Height of the hub in meters.
d_rotor	Diameter of the rotor in meters.
wind_conv_type	Name of the wind converter according to the list in the csv file

For a PV module, the parameters listed in Table 4.5 can be used.

Table 4.5 List of parameters with examples for the yield of a PV module (oemof developing group 2015)

Azimuth	Azimuth angle of the pv module in degree.
Tilt	Tilt angle of the pv module in degrees.
module_name	According to the sandia module library (cf. Sect. 4.2).
Albedo	Albedo value.

OEMOF makes data for yields of PV modules and wind power plants available to modelers. These datasets contain the parameters listed above. But modelers can also enter data they have gathered themselves into their model. In order to initialize a solar module or a wind power plant, the user must create a factsheet in form of an excel sheet or other spreadsheet format, which provides the data required to compute the yield. One key feature of this is the name of the make. This name must be given as the first parameter on the factsheet.

The following source code can be used to define a solar module or wind power plant for modeling in OEMOF:

```
# The following code generates a wind turbine and a pv module object
# Each object is assigned to a model (SimpleWindModel and PvlibBased)
# of the class 'WindPowerPlant' or 'Photovoltaic'.
# These models are used to compute the feedin data.
# The parameter set is also passed to the objects. This is done by means of
# '**my_parameter_set'. The two asterisks in front transfer a dictionary
# as keyed-parameter.
my_wind_turbine = plants.WindPowerPlant(model=SimpleWindModel,
                                        **my_parameter_set)
my_pv_module = plants.Photovoltaic(model=PvlibBased, **my_parameter_set)
```

The keyed-parameter mentioned in the code excerpt above is the parameter of a function that is assigned not by means of its position but by means of the name that it is assigned to. An example for this is the name model=SimpleWindModel. If arguments for this name are stored in a dictionary (as a name-value pair) these values can be assigned to the relevant keyed-parameters, if the dictionary is preceded by two asterisks '**'. Without the asteriks each value would have to be assigned separately to the named entity.

The library *feedinlib* computes the input data (time series/load profiles) for PV plants and wind power plants from weather data sets. The electricity yield is computed based on the dataset of the specified plants. The *feedinlib* library contains two separate libraries, the *pvlib* and the *windpowerlib*. *Feedinlib* makes a model for wind power and one for PV power available to modelers. But of course, it is also possible to compute the yield based on your own data and to make use of your own model.

If modelers want to use their own models, the models must be fed with the relevant data.

```
own_wind_model = models.YourWindModelClass(required=[parameter1,
                                           parameter2])
own_pv_model = models.YourPVModelClass()

your_wind_turbine = plants.WindPowerPlant(model=own_wind_model,
                                          **your_parameter_set)
your_pv_module = plants.Photovoltaic(model=own_pv_model,
                                     **your_parameter_set)

feedin_series_wp1 = your_wind_turbine.feedin(data=my_weather_df, number=5)
feedin_series_pv1 = your_pv_module.feedin(data=my_weather_df)  # One Module
```

Using *feedinlib* OEMOF intitializes an encapsulated weather object. A weather object requires a set of weather data as well as some other meta-data, such as its location. In order to be able to initialize the object, first the class *FeedinWeather* is created. The modeler's weather data are then passed to this class:

```
my_weather_a = weather.FeedinWeather(
    data=my_weather_pandas_DataFrame,
    timezone='Continent/City',
    latitude=x,
    longitude=y,
    data_heigth=coastDat2
    )
```

As a *timezone* Europe/Berlin or Australia/Sydney could be selected. The attributes *latitude* and *longitude* receive floating point values. In OEMOF not all parameters for the weather data must be filled, for example *latitude* and *longitude* do not have to be specified. The parameter *timezone* is also not required, if a complete time index (full time index) with a timezone was provided.

The item *coastDat2* is a special dictionary for data heights. If the optimization model is to compute yields, the PV model requires information on radiation, temperature, wind speed and roughness length. The data for the *data_height* dicionary must be written as shown in the excerpt below:

```
coastDat2 = {
    'dhi': 0,
    'dirhi': 0,
    'pressure': 0,
```

```
'temp_air': 2,
'v_wind': 10,
'z0': 0}
```

If modelers use their own dictionary with their own column headings, these headings must be overwritten with the terms used in the *feedinlib*. This can be done using the following commands:

```
name_dc = {
    'my diffuse horizontal radiation': 'dhi',
    'my direct horizontal radiation': 'dirhi',
    'my pressure data set': 'pressure',
    'my ambient temperature': 'temp_air',
    'my wind speed': 'v_wind',
    'my roughness length': 'z0'}

my_weather_DataFrame.rename(columns=name_dc)
```

If the *feedinlib* is used together with the OEMOF package, the data for the *time series* must be passed on to the optimization model. For this the weather object created must be passed on to the own model. If only the weather object is passed on, the electrical capacity for a wind turbine can be calculated. For computing data of just one PV module, the following code excerpt can be used:

```
feedin_series_pv1 = my_pv_module.feedin(weather=my_weather_df)
```

It is also possible to compute data for more than one wind turbine or PV module. In order to do this, the parameters *number* and *installed capacity* can be defined. For PV modules, the parameters *number*, *peak_power* and *area* can be used. To compute data of several wind turbines, the following source code can be used:

```
feedin_series_wp1 = your_wind_turbine.feedin(data=my_weather_df,
number=5)
```

The calculations for wind power and PV modules are based on the following mathematical descriptions.

Computation basis for PV modules

The computation method for the PV module can be found in the pvlib library at: http://pvlib-python.readthedocs.io/en/latest/ (Krien 2017).

For a PV module the solar position and the incident angle of the sun are relevant. The incident angle can be calculated based on the solar position and the positioning of the PV module [http://wholmgren-pvlib-python-new.readthedocs.io/en/doc-reorg2/generated/pvlib.irradiance.html (GitHub o.J.). In order to calculate the solar position, *feedinlib* provides the ephemeris model of the pvlib (pvlib.solarposition.

get_solarposition). Many weather databases also provide median hourly values for the solar position, which can be used for the calculations. For this reason, the model included in *feedinlib* also calculates hourly median solar position values based on 5-minute values (Krien 2017). In order to ensure that no values that cannot be displayed are returned (nan-values) any values that are lower than 0° or higher than 90° are discarded.

The computation model in *pvlib* uses the direct normal irradiation value: dni. This value is calculated in *feedinlib* using a geometric calculation (cf. Eq. 4.1):

$$dni = \frac{dirhi}{\sin\left(\frac{\pi}{2} - \alpha_{\text{zenith}}\right)} \tag{4.1}$$

where:

dni direct horizontal irradiation, this value is provided by the weather dataset
α_{zenith} solar zenith angle

The *dni* value can vary considerably if there are even minor discrepancies in the solar position calculations provided in the weather model or fed into the weather model from own datasets. But this is normally only the case, if the solar angle (α_{zenith}) is more than 88° as this means that the denominator in Eq. 4.1 is very small. Following the assumption that irradiation is negligible so close to the zenith and that the irradiation effect is low for in-plane irradiance, for α_{zenith} greater than 88° the direct horizontal irradiation *dirhi* can be considered equal to *dni* (cf. Eq. 4.2).

$$dni = dirhi \tag{4.2}$$

In-plane Irradiance is defined as the sunlight that illuminates tilt plane surface (Tamura 2003).

For calculating diffuse radiation from the sky, *feedinlib* uses the Perez model. The following methods are used by *pvlib* to calculate this (Krien 2017):

- pvlib.irradiance.extraradiation
- pvlib.atmosphere.relativeairmass
- pvlib.irradiance.perez
- pvlib.irradiance.grounddiffuse
- pvlib.irradiance.globalinplane.

The term *pvlib.irradiance.grounddiffuse* is used to calculate ground reflection. Via the computation *global in-plane* (*pvlib.irradiance.globalinplane*) the different light refractions are added up.

The electricity output of a PV module is expressed via the term *pvlib.pvsystem. sapm*. To calculate the electricity output *feedinlib* uses the Sandia PV Array Performance Model (SAPM). The electricity output can only be calculated once the temperature of the solar module has been calculated (*pvlib.pvsystem.sapm_cell-temp*). The temperature of the solar module is an important factor for SAPM.

Computation basis for wind turbines

The basis for calculating the yield of a wind turbine is simple (Krien 2017). The main factor is the cp value (coefficient of power) of one specific wind turbine. The cp data for a wind turbine can be requested from the manufacturer. The values are listed as a sequence in separate steps with increments of 0.5/1 m/s wind speed. For the feedinlib calculation, cp values plotted as continuous graph over wind speed is decisive. In order to generate this information *feedinlib* carries out a linear literpolation of cp values (cf. Eq. 4.3),

$$P_{wpp} = \frac{1}{8} \cdot \rho_{air,hup} \cdot d_{rotor}^2 \cdot \pi \cdot v_{wind,hub}^3 \cdot cp\left(v_{wind,hub}\right) \qquad (4.3)$$

where:

d_{rotor} diameter of the rotor in (m)

$\rho_{air,hup}$ air density at hub height, estimated temperature 6.5 K and a pressure gradient of $-1/8$ hPa/m

$v_{wind,hub}$ wind speed at hub height

cp   cp-values against the wind speed

The wind speed at hub height is calculated by means of Eq. 4.4,

$$v_{wind,hub} = v_{wind,data} \cdot \frac{\ln\left(\frac{h_{hub}}{z_0}\right)}{\ln\left(\frac{h_{wind,data}}{z_0}\right)} \qquad (4.4)$$

where:

$v_{wind,hub}$ wind speed at the height of the weather model or measurement

h_{hub} height of the hub

$h_{wind,data}$ height of the wind speed measurement or height of the wind speed within the weather model

By means of Eq. 4.4: a logarithmic wind profile is computed. The temperature at hub height is calculated via Eq. 4.5:

$$T_{Hub} = T_{air,data} - 0.0065 \cdot \left(h_{hub} - h_{T,data}\right) \qquad (4.5)$$

where

$T_{air,data}$ temperature at the height of the weather model or measurement

h_{hub} height of the hub

$h_{T,data}$ height of temperature measurement or the height of the temperature within the weather model

Once the temperature at hub height has been calculated, the next step is to calculate the air density (cf. Eq. 4.6),

$$\rho_{\text{air,hub}} = \frac{\left(\frac{p_{\text{data}}}{100} - \left(h_{\text{hub}} - h_{\text{p,data}}\right) \cdot \frac{1}{8}\right)}{\left(2.8706 \cdot T_{\text{hub}}\right)} \tag{4.6}$$

where

p_{data}	pressure at the height of the weather model or measurement
T_{hub}	temperature of the air at hub height
h_{hub}	height of the hub
$h_{\text{p,data}}$	height of pressure measurement or the height of pressure within the weather model

More information about the programming code in feedinlib

The *feedinlib* library contains the following three modules:

- *models.py*
  This can be found at: https://github.com/oemof/feedinlib/blob/dev/feedinlib/ models.py
- *powerplants.py*
  This can be found at: https://github.com/oemof/feedinlib/blob/dev/feedinlib/ powerplants.py
- *weather.py*
  This can be found at: https://github.com/oemof/feedinlib/blob/dev/feedinlib/ weather.py

Let us start with the module *models.py*. Several classes and their methods are defined in this module. The first class is the base class *Base(ABC)* for the *feedinlib* models:

```
class Base(ABC):
```

The class *PvlibBased(Base)* is required for defining the output of PV modules in a model:

```
class PvlibBased(Base)
```

This is where the calculations based on the *pvlib* library are carried out. The following code executes the command to import pvlib models:

```
from feedinlib import models
pv_model = models.PvlibBased()
```

This class contains the method *def feedin(self, **Kwargs)*, by means of which the time series for the PV module are read. These are returned as Pandas series for a given PV model.

The next method *def solarposition_hourly_mean(self, location, data, **kwargs)* calculates the mean hourly solar position. This is calculated by means of angles across the horizon. Several parameters are required for this method, which are listed in Table 4.6.

Table 4.6 Parameters for the method *def solarposition_hourly_mean(self, location, data, **kwargs)* (GitHub 2018b)

Parameter	Term	Description
Location	pvlib.location. Location	A pvlib location object containing the longitude, latitude and the timezone of the location.
Data	pandas. DataFrame	Containing the time index of the location.
Method	string, optional	Method to calculate the position of the sun according to the methods provided by the pvlib function (default: 'ephemeris') 'pvlib.solarposition.get_solarposition'.
Freq	string, optional	The time interval to create the hourly mean (default: '5Min').

The *pandas.DataFrame* provides the additional parameters azimuth, zenith and elevation. The source code below can be used to calculate and return the solar position for a defined time:

```
data_5min = pd.DataFrame(
    index=pd.date_range(data.index[0],
                        periods=data.shape[0]*12, freq='5Min',
                        tz=kwargs['weather'].timezone))

data_5min = pvlib.solarposition.get_solarposition(
time=data_5min.index, latitude=location.latitude,
longitude=location.longitude, method='ephemeris')

return pd.concat(
[data, data_5min.clip_lower(0).resample('H').mean()],
axis=1, join='inner')
```

Via the method *def solarposition(self, location, data, **kwargs)* the solar position is determined based on the time index. This is carried out by the following sequence:

```
import pvlib
import pandas as pd
```

```
from feedinlib import models
loc = pvlib.location.Location(52, 13, 'Europe/Berlin')
pvmodel = models.PvlibBased()
data = pd.DataFrame(index=pd.date_range(pd.datetime(2010, 1, 1, 0),
                        periods=8760, freq='H', tz=loc.tz))
elevation = pvmodel.solarposition(loc, data).elevation
print(round(elevation[12], 3))
14.968
```

```
return pd.concat(
                [data, pvlib.solarposition.get_solarposition(
                time=data.index, latitude=location.latitude,
                longitude=location.longitude,
                method=kwargs.get('method', 'ephemeris'))],
                axis=1, join='inner')
```

The incidence angle is calculated by *pvlib aoi function* by means of the method *def angle_of_incidence(self, data, **kwargs)*. To run this method the parameters listed in Table 4.7 are required.

Table 4.7 Parameters for the method *def angle_of_incidence(self, data, **kwargs)* (GitHub 2017c)

Parameter	Term	Description
Data	pandas. DataFrame	Containing the timeseries of the azimuth and zenith angle.
Tilt	float	Tilt angle of the pv module (horizontal = 0°).
Azimuth	float	Azimuth angle of the pv module (south = 180°).

The angle of incidence is returned in degrees as pandas.series via this command:

```
return pvlib.irradiance.aoi(
            solar_azimuth=data['azimuth'], solar_zenith=data['zenith'],
            surface_tilt=self.powerplant.tilt,
            surface_azimuth=self.powerplant.azimuth)
```

Via the method *def global_in_plane_irradiation(self, data, **kwargs)* the global irradiaton on the tilted surface is calculated. This method takes into account the direct and diffuse irradiation, the incident angle and the orientation of the surface in order to calculate global irradiation in plane. To calculate these, the methods *pvlib. irradiance.globalinplane function* as well as other methods of the modules *pvlib. atmosphere* and the *pvlib.solarposition* are used to generate the input parameters for the *globalinplane function*.

The parameters used for this method are listed in Table 4.8.

Table 4.8 Parameter for the method *def global_in_plane_irradiation(self, data, **kwargs)* (GitHub 2017c)

Parameter	Term	Description
Data	pandas. DataFrame	Containing the time index of the location and columns with the following timeseries: (dirhi, dhi, zenith, azimuth, aoi).
Tilt	float	Tilt angle of the pv module (horizontal = 0°).
Azimuth	float	Azimuth angle of the pv module (south = 180°).
Albedo	float	Albedo factor around the module.

The *Pandas.DataFrame* provides further columns with the headings poa_global, poa_diffuse und poa_direct.

Next follow some methods for calculating key factors for the computation (GitHub 2017c):

- Determine the extraterrestrial radiation:

```
data['dni_extra'] =
pvlib.irradiance.extraradiation(datetime_or_doy=data.index.dayofyear)
```

- Determine the relative air mass:

```
data['airmass'] = pvlib.atmosphere.relativeairmass(data['zenith'])
```

- Determine direct normal irradiation:

```
data['dni'] = (data['dirhi']) / np.sin(np.radians(90 - data['zenith']))
```

- The next line provides the instructions of how to act if the αzenith is greater than 88° as in that case the direct horizontal irradiation *dirhi* is equated with the direct radiation *dni*:

```
data['dni'][data['zenith'] > 88] = data['dirhi']
```

- Determine the sky diffuse irradiation in plane with model of Perez:

```
        data['poa_sky_diffuse'] = pvlib.irradiance.perez(
            surface_tilt=self.powerplant.tilt,
            surface_azimuth=self.powerplant.azimuth,
            dhi=data['dhi'],
            dni=data['dni'],
            dni_extra=data['dni_extra'],
            solar_zenith=data['zenith'],
            solar_azimuth=data['azimuth'],
            airmass=data['airmass'])
```

– Set NaN values to zero:

```
data['poa_sky_diffuse'][
    pd.isnull(data['poa_sky_diffuse'])] = 0
```

– Determine the diffuse irradiation from ground reflection in plane:

```
data['poa_ground_diffuse'] = pvlib.irradiance.grounddiffuse(
    ghi=data['dirhi'] + data['dhi'],
    albedo=self.powerplant.albedo,
    surface_tilt=self.powerplant.tilt)
```

– Determine total in-plane irradiance:

```
data = pd.concat(
    [data, pvlib.irradiance.globalinplane(
        aoi=data['aoi'],
        dni=data['dni'],
        poa_sky_diffuse=data['poa_sky_diffuse'],
        poa_ground_diffuse=data['poa_ground_diffuse'])],
    axis=1, join='inner')

return data
```

The following command fetches the module data from the Sandia Module Library:

```
def fetch_module_data(self, lib='sandia-modules', **kwargs)
```

For creating the data output from the PV module the following command is used:

```
def pv_module_output(self, data, **kwargs)
```

For this the method *pvlib.pvsystem.sapm* from the *pvlib* is used, the relevant parameters are listed in Table 4.9.

Table 4.9 Parameters for the method *pv_module_output(self, data, **kwargs)* (GitHub 2017c)

Parameter	Bezeichnung	Description
module_name	string	Name of a pv module from the sam.nrel database.
Data	pandas. DataFrame	Containing the time index of the location and columns with the following timeseries: (temp_air [K], v_wind, poa_global, poa_diffuse, poa_direct, airmass, aoi).
Method	string, optional	Method to calculate the position of the sun according to the methods provided by the pvlib function (default: 'ephemeris') 'pvlib.solarposition.get_solarposition'.

The Pandas DataFrame (pandas.DataFrame) provides two additional columns with two further parameters *p_pv_norm* and *p_pv_norm_area*. By means of the following method the overall output of a PV module can be returned (GitHub 2017c):

```
def get_pv_power_output(self, **kwargs)
```

The next section provides methods and functions for the class *SimpleWindTurbine(Base)*:

```
class SimpleWindTurbine(Base)
```

In order to call this class up in your own application, the following code can be used:

```
from feedinlib import models
required_ls = ['h_hub', 'd_rotor', 'wind_conv_type', 'data_height']
wind_model = models.SimpleWindTurbine(required=required_ls)
```

The following parameters are required for defining the wind power model (cf. Table 4.10).

Table 4.10 Parameters for the model *SimpleWindTurbine* (GitHub 2017c)

Parameter	Data type	Description
h_hub	float	Height of the hub of the wind turbine.
d_rotor	float	Diameter of the rotor (m).
wind_conv_type	string	Name of the wind converter type. Use self.get_wind_pp_types() to see a list of all possible wind converters.

The next module *powerplants.py* contains the source code for initializing the classes (GitHub 2017d):

– class *Base(ABC)*
  This is the base class for power plants in *feedinlib*
– class *Photovoltaic(Base)*
– class *WindPowerPlant(Base)*

As these classes do not require any specific methods, they will not be further discussed here. Further information can be found at (GitHub 2017c).

The final module *weather.py* contains only one class:

– class *FeedinWeather*

This class contains all meta-data and further information required for the weather datasets.

The following parameters (cf. Table 4.11) are relevant for this class.

Table 4.11   Parameters for *FeedinWeather* (GitHub 2017e)

Parameter	Data type	Description
Data	pandas. DataFrame, optional	Containing the time series of the different parameters as columns.
Timezone	string, optional	Containing the name of the time zone using the naming of the IANA (Internet Assigned Numbers Authority) time zone database.
Longitude	float, optional	Longitude of the location of the weather data.
Latitude	float, optional	Latitude of the location of the weather data.
Geometry	shapely.geometry object	polygon or point representing the zone of the weather data.
data_height	dictionary, optional	Containing the heights of the weather measurements or weather model in meters with the keys of the data parameter.
Name	string	Name of the weather data object.

4.1.8   Oemof-Demandlib

The library *demandlib* provides electricity and heat load profiles for energy consumption. These are generated by means of data provided to users by the German Federal Association of Energy and Water Management (BDEW). Modelers can use the load profiles of the BDEW and adapt them to calculate their annual demand.

The central module for importing load profiles in OEMOF is *bdew.py*. This module provides two classes, one for the electricity load profile and one for the heat load profile for buildings (oemof developing group 2016):

- *class demandlib.bdew.ElecSlp(year, seasons=None, holidays=None)*
  Based on the BDEW method this class generates standardized electricity load profiles.
- *class demandlib.bdew.HeatBuilding(df_index, **kwargs)*
  This class creates heat load profiles for buildings.

The first class *ElecSlp* has the following attributes (cf. Table 4.12).

Table 4.12 Attribute of the class *demandlib.bdew.ElecSlp(year, seasons=None, holidays=None)* (oemof developing group 2016)

Attribute	Data type	Description
datapath	string	Path to the csv files containing the load profile data.
date_time_index	pandas. DateTimeIndex	Time range for and frequency for the profile.

The following Table 4.13 lists the parameters required to calculate load profiles.

Table 4.13 Parameters of the class *demandlib.bdew.ElecSlp(year, seasons=None, holidays=None)* (oemof developing group 2016)

Parameter	Data type	Description
Year	Integer	Year of the demand series.
Optional parameters		
Seasons	Dictionary	Describing the time ranges for summer, winter and transition periods.
Holidays	Dictionary or list	The keys of the dictionary or the items of the list should be datetime objects of the days that are holidays.

There are three more methods that are used to calculate the load profiles (cf. Table 4.14).

Table 4.14 Methods of the class *demandlib.bdew.ElecSlp(year, seasons=None, holidays=None)* (oemof developing group 2016)

Method	Description
all_load_profiles(time_df, holidays=None)	This is used to upload the profiles.
create_bdew_load_profiles(dt_index, slp_types, holidays=None)	Calculates the hourly electricity load profile in MWh/h of a region.
get_profile(ann_el_demand_per_sector)	Get the profiles for the given annual demand.

The attribute *ann_el_demand_per_sector* in the method *get_profile* is a dictionary containing annual summed values. The result is a table with all load profiles in form of a *pandas.DataFrame*.

For the class *HeatBuilding* there is only one parameter (cf. Table 4.15).

Table 4.15 Parameter for the class *demandlib.bdew.HeatBuilding(df_index, **kwargs)* (oemof developing group 2016)

Parameter	Data type	Description
Year	integer	Year for which the profile is created.

There are several attributes for this class, these are listed in Table 4.16.

Table 4.16 Attributes for the class *demandlib.bdew.HeatBuilding(df_index, **kwargs)* (oemof developing group 2016)

Attribute	Data type	Description
Datapath	string	Path to the bdew basic data files (csv).
Temperature	pandas. Series	Series containing hourly temperature data.
annual_heat_demand	float	Annual heat demand of building in kWh.
building_class	integer	Class of building according to bdew classification possible numbers are: 1–11.
shlp_type	string	Type of standardized heat load profile according to bdew possible types are: GMF, GPD, GHD, GWA, GGB, EFH, GKO, MFH, GBD, GBA, GMK, GBH, GGA, GHA.
wind_class	integer	Wind classification for building location (0 = not windy or 1 = windy).
ww_incl	boolean	Decider whether warm water load is included in the heat load profile.

The class *HeatBuilding* also has several methods (cf. Table 4.17).

Table 4.17 Methods of the class *demandlib.bdew.HeatBuilding(df_index, **kwargs)* (oemof developing group 2016)

Method	Description
get_bdew_profile()	Calculation of the hourly heat demand using the bdew-equations.
get_sf_values (filename='shlp_hour_factors.csv')	Determine the h-values. Very important is the parameter*filename* (stri ng)—name of file where sigmoid factors are stored.
get_sigmoid_parameters (filename='shlp_sigmoid_factors. csv')	Retrieve the sigmoid parameters from csv-files. Again important, the parameter *filename* (string)—name of file where sigmoid factors are stored.
get_temperature_interval()	Appoints the corresponding temperature interval to each temperature in the temperature vector.
get_weekday_parameters (filename='shlp_weekday_factors. csv')	Retrieve the weekday parameter from csv-file, with the parameter *filename*.
weighted_temperature (how='geometric_series')	A new temperature vector is generated containing a multi-day average temperature as needed in the load profile function. The parameter here is the *how* (string) —string which type to return (*geometric_series* or *mean*).

For generating a series of average temperatures Eq. 4.7 can be used.

$$T = \frac{T_D + 0.5 \cdot T_{D-1} + 0.25 \cdot T_{D-2} + 0.125 \cdot T_{D-3}}{1 + 0.5 + 0.25 + 0.125} \tag{4.7}$$

where

T_D average temperature on the present day
T_{D-i} average temperature on the day $- i$

4.2   Procurement of Data

The procurement of relevant data is a key element of successful modeling. Creating the perfect model will not achieve its goal if the relevant data cannot be obtained. When gathering data, there are two aspects:

– How current are the data?
– How can the data be accumulated?

Accumulation can be done both by sorting according to date and time and by sorting according to location. Depending on the question that is to be answered, minute data, quarter of an hour data or even data per second may be required. The location scaling will also depend on the question. The modeling question will also determine the balance limit that will be drawn. This balance boundary will also determine, if data are to be gathered for an entire region, a whole country or only one municipality or even just one residential area. The smaller the overall scale of the model, the more detailed the data gathering and the data analysis.

In addition to time series and location data, data will be required to cover:

- Electricity load profiles for:

  – buildings
  – manufacturing

- Heat load profiles for:

  – buildings, heating, hot water
  – manufacturing, heating, hot water
  – manufacturing, process heat

- Vehicle mobility with the parameters:

  – fuel/energy content
  – travel distance per fuel tank/energy content

- Plant technology
- Fuels

- Energy grids (gas, electricity, district heating)
- Potential energy, such as waste to be used for biogas, potential energy options for power-to-heat, solar arrays on rooftops or green fields.

Before they get started, modelers must ask themselves the question where they might get the data from. Data are a valuable commodity and obtaining accurate and relevant data can therefore be very expensive. But there is also a certain current trend to offer data for free. For example, the Research Institute GmbH in Jülich has started a project on 1st January 2017 with the title *Main project: harmonization and development of methods for a spatial and temporal resolution of energy demands (DemandRegio)* with the subproject entitled *Indicators*. This project is to be completed by December 31, 2019. One objective of this project is to develop a harmonized and transparent approach to regionalize the electricity and gas demand with high temporal resolution. The gathered data is to be made available to interested parties as an open data initiative. By harmonizing data and thus reducing data uncertainty it will be easier to compare different modeling and scenario results. This will also mean that results of such modeling exercises can be published in the knowledge that they offer a certain reliability.

Another project is called *OPSD project* (https://open-power-system-data.org/). This project makes its Open Power System Data Platform available to users. One key aspect of this project is to harmonize previously published data and offer it in a unified format, so that the data can be reused and applied to different models.

Further data sources include:

- Studies
  It is very important to ensure that the study uses current data. Another relevant aspect is to look at the method that was used for gathering data in the study. This is particularly important if users plan to combine data from different studies. Studies are conducted by associations such as the German Federal Association of Energy and Water Management (BDEW), The German Federal Association of Renewable Energies (BEE) or the German Federal Association for Wind Power (BWE). Other studies may have been commissioned by corporations, such as Agora Energiewende (Agora Energy Transition). Other studies may be conducted by consultancy companies acting on behalf of associations or political institutions.
- Energy providers
  Energy providers supply gas, electricity and heat. Some of these providers make their data available online. But in most cases, it is still necessary to contact the providers and ask for the data in the required format and for the required period and location. Figures 4.7 and 4.8 show a map of all grid providers in Germany. As can be seen, the grid providers differ depending on the voltage (high, medium and low voltage grids).

Fig. 4.7 Map of energy providers in a medium voltage grid (ene't GmbH 2017b)

Fig. 4.8 Map of energy providers in the high voltage grid 2017 (ene't GmbH 2017a)

The grid providers for high voltage grids act supranationally, whereas there are big regional concentrations for the medium voltage grids. This means that in order to obtain data for load profiles, different companies will have to be contacted for the medium voltage and the high voltage electricity providers.

- Databases
  One relevant database is the energy database (http://www.energie-datenbank.eu/), which provides technical data on technology components for wind energy, solar energy and heat supply. The Fraunhofer Institute of Solar Energy (ISE) makes data and graphs available on their website https://www.energy-charts.de/index_de.htm. The EUROSTAT data base (http://ec.europa.eu/eurostat/de/web/energy/data/database) provides several free statistics and statistics tools. The website of the Agora Energy Transition corporation provides the tool Agorameter (https://www.agora-energiewende.de/de/themen/-agothem-/Produkt/produkt/76/Agorameter/) which can help create graphs for load profiles.
- GIS Maps
  Most maps provided by GIS require payment. Still, using this technology can be worthwhile for use in the energy sector.
- Building stocks and industry
  Data about buildings and inhabitants for a certain region can be obtained from the German Federal Institute for Statistics. The database Eurostat also provides statistical data. But this data tend to be accumulated for the whole of Germany or the whole of Europe.
- Weather data
  German weather data are available via the German Weather Service (DWD) (https://www.dwd.de/DE/Home/home_node.html). The German website https://www.wetteronline.de/rueckblick?gid=euro also provides weather data for the whole of Europe.

Weather data for the U.S. and international can be found on https://www.ncdc.noaa.gov/.

Global weather data that have been made available in OEMOF in the *feedinlib* library have been extracted from the coastDat2 database (https://www.earth-syst-sci-data.net/6/147/2014/) of the Institute of Coastal Research, Helmholtz-Zentrum Geesthacht, Geesthacht, Germany. This data were fed into *feedinlib*, but are only available via license and are not available as a whole. Requests for access to this data should be directed at the Helmholtz Institute in Geesthacht.

OEMOF also provides data for PV modules obtained from Sandia Laboratories at https://sam.nrel.gov/sites/default/files/sam-library-sandia-modules-2015-6-30.csv   or https://sam.nrel.gov/sites/default/files/sam-library-cec-modules-2015-6-30.csv. The cp values for wind power plants have been provided by the Reiner Lemoine Institute. This can be downloaded from http://vernetzen.uni-flensburg.de/~git/cp_values.csv.

Once the data have been gathered, they are entered into the model as a csv file (cf. Fig. 4.9). The column headings must be prepared to fit into the OEMOF system by giving relevant instructions in the source code.

Fig. 4.9 Example for the feed-in of a CSV file, here a representation of energy sources

The selection of data must be based on the problem set in the optimization model.

4.3   The Program Spyder

Programming languages have their own semantics, which will be represented in a certain way in the browser. Some browsers will indent source code and color some terms in order to make the code more readable, others not. And both modelers and OEMOF developers must continually work on testing and debugging their code. Furthermore, a graphic representation of results will be required at the end.

These functionalities are served by the Spyder software. This is another open source system of programs, which is part of the OEMOF installation. Spyder was developed in Python (The Spyder Project o.J.). It is a Scientific Python Development Environment. By using IPython (enhanced interactive Python interpreter) and popular Python libraries such as NumPy (linear algebra), SciPy (signal and image processing) or matplotlib (interactive 2D/3D plotting) it provides a numerical computing environment.

4.3.1   Spyder Editor

The programming code in Spyder is formulated via the editor. For this the Spyder work surface is divided into several windows each fullfilling a different task. On the left side is the text editor for writing programming code (cf. Fig. 4.10).

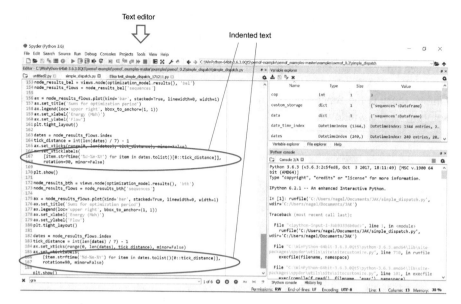

Fig. 4.10 Work surface of Spyder, text editor on the left side (The Spyder Project o.J.)

The colors used in the code are significant. (cf. Fig. 4.11). The color blue is used for keywords and green for strings. But modelers can adapt the color code to their requirements.

Fig. 4.11 Spyder syntax coloring

The editor allows modelers to work in different languages (multi-language editor). In addition to the helpful color coding, the code is also analyzed in real-time (cf. Fig. 4.12). Furthermore, Spyder auto-completes terms, classes and methods and provides tips for methods (calltips) and offers go-to-definition features (powered by rope) (The Spyder Project o.J.). Spyder has many other useful functionalities. One of these is the function/class browser and the option of horizontal or vertical splitting of the windows.

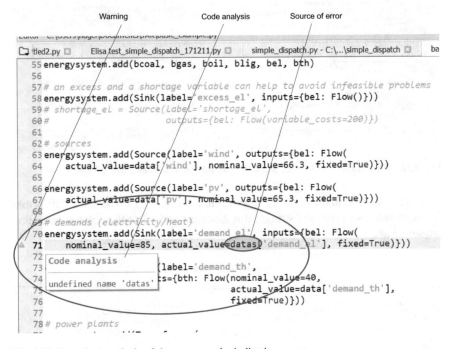

Fig. 4.12 Real-time analysis of the source code, indicating errors

4.3.2   *Variable Explorer*

The windows in Spyder can be split vertically or horizontally (cf. Fig. 4.13). The top window on the left lists the variables in the variable explorer by name and describes their function. The window on the lower left side contains the IPython console. The console lists the steps already completed in the modeling process, returns results and flags up errors.

In the lower half on the left Fig. 4.13 shows the optimization results depicted as a graph. To obtain this visualization commands must be given, as described in Sect. 4.1.6.

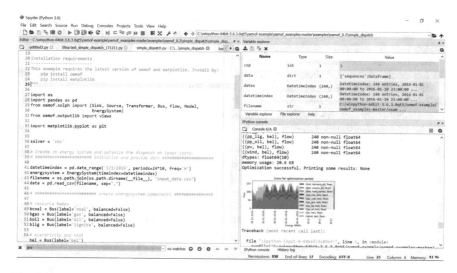

Fig. 4.13 Vertical window layout in Spyder

As a final point, we want to look at visualization options once more. The visualization described in Sect. 4.1.6 is only possible by using the program PyGraphviz. But OEMOF also provides visualization as a network structure. This type of visualization can be carried out in Spyder by means of the IPython Console.

For more detailed information on Spyder as a program system we recommend see (The Spyder Project o. J.).

References

ene't GmbH: Karte der Stromnetzbetreiber Hochspannung (2017a). https://www.enet.eu/assets/media/images/karten/2017/Karte-der-Stromnetzbetreiber-Hochspannung-2017.jpg. Accessed on 17 Jan 2018

ene't GmbH: Karte der Stromnetzbetreiber Mittelspannung (2017b). https://www.enet.eu/assets/media/images/karten/2017/Karte-der-Stromnetzbetreiber-Mittelspannung-2017.jpg. Accessed on 17 Jan 2018

Gabler Wirtschaftslexikon: WACC. Springer Gabler (2018). https://wirtschaftslexikon.gabler.de/definition/wacc-50279/version-273501. Accessed on 28 Apr 2018

Gaudschau, E. et al. (Authors), Fraktion BÜNDNIS 90/DIE GRÜNEN im Brandenburger Landtag (Ed.): Kohleausteig à la Brandenburg: Funktioniert die Energiestrategie 2013 der Brandenburger Landesregierung? Studie des Reiner Lemoine Instituts gGmbH (RLI) und der Hochschule für Technik und Wirtschaft (HTW Berlin) im Auftrag der Fraktion BÜNDNIS 90/DIE GRÜNEN im Brandenburger Landtag Potsdam, (2017). https://www.gruene-fraktion-brandenburg.de/fileadmin/ltf_brandenburg/Dokumente/Publikationen/Untersuchungen_zur_Energiestrategie_2030.pdf, Accessed on 20 Jan 2018

GitHub Inc.: oemof/oemof/network.py (2017a). https://github.com/oemof/oemof/blob/dev/oemof/energy_system.py. Accessed on 14 Dec 2017

GitHub Inc.: oemof/oemof/solph/constraints.py (2017b). https://github.com/oemof/oemof/blob/dev/oemof/tools/helpers.py. Accessed on 19 Dec 2017

GitHub Inc.: feedinlib/feedinlib/models.py (2017c). https://github.com/oemof/feedinlib/blob/dev/feedinlib/models.py. Accessed on 15 Jan 2018

GitHub Inc.: feedinlib/feedinlib/powerplants.py (2017d). https://github.com/oemof/feedinlib/blob/dev/feedinlib/powerplants.py. Accessed on 15 Jan 2018

GitHub Inc.: feedinlib/feedinlib/weather.py (2017e). https://github.com/oemof/feedinlib/blob/dev/feedinlib/weather.py. Accessed on 15 Jan 2018

GitHub Inc.: oemof/oemof/outputlib/ (2018a). https://github.com/oemof/oemof/blob/dev/oemof/outputlib. Accessed on 02 Feb 2018

GitHub Inc.: feedinlib/feedinlib/models.py (2018b). https://github.com/oemof/feedinlib/blob/dev/feedinlib/models.py. Accessed on 02 Feb 2018

GitHub Inc.: pvlib.irradiance. o.J. http://wholmgren-pvlib-python-new.readthedocs.io/en/doc-reorg2/generated/pvlib.irradiance.htm . Accessed on 02 Feb 2018

Graphviz: Graph Visualization. o.J. http://www.graphviz.org/ . Accessed on 03 Feb 2018

Klein, B.: Properties vs. Getters and Setters. o.J.-a. https://www.python-course.eu/python3_properties.php . Accessed on 20 Jan 2018

Klein, B.: Python-Kurs: Formatierte Ausgabe. o.J.-b. https://www.python-kurs.eu/python3_formatierte_ausgabe.php . Accessed on 01 Mar 2018

Klein, B.: Python-Kurs: for-Schleifen. o.J.-c. https://www.python-kurs.eu/for-schleife.php . Accessed on 01 Mar 2018

Klein, B.: Python-Kurs: Properties als Ersatz für Getter und Setter Properties. o.J.-d. https://www.python-kurs.eu/python3_properties.php . Accessed on 01 Mar 2018

Krien, U. et al.: feedinlib Documentation. Release beta (2017). https://media.readthedocs.org/pdf/feedinlib/v0.0.10/feedinlib.pdf. Accessed on 14 Jan 2018

oemof-developer-group: oemof-outputlib (2014). http://oemof.readthedocs.io/en/stable/oemof_outputlib.html#creating-an-input-output-plot-for-buses. Accessed on 07 Mar 2018

oemof developing group: Getting started (2015). http://pythonhosted.org/feedinlib/getting_started.html#initialise-your-turbine-or-module. Accessed on 13 Jan 2018

oemof-developer-group: Getting started (2016). http://demandlib.readthedocs.io/en/latest/getting_started.html#introduction. Accessed on 21 Nov 2017

oemof-Team: oemof Documentation (2017a). https://media.readthedocs.org/pdf/oemof/latest/oemof.pdf. Accessed on 20 Jan 2018

oemof-Team: oemof Documentation (2017b). https://media.readthedocs.org/pdf/oemof/stable/oemof.pdf. Accessed on 12 Jan 2017

oemof-developer-group: oemof-outputlib (2018a). https://github.com/oemof/oemof_visio/blob/master/oemof_visio/plot.py. Accessed on 07 Mar 2018

oemof-developer-group: oemof_examples/examples/oemof_0.2/plotting_examples/storage_investment_plot.py (2018b). https://github.com/oemof/oemof_examples/blob/master/examples/oemof_0.2/plotting_examples/storage_investment_plot.py. Accessed on 07 Mar 2018

pandas: pandas: powerful Python data analysis toolkit (2017a). http://pandas.pydata.org/pandas-docs/stable/. Accessed on 12 Dec 2017

pandas: pandas.date_range (2017b). http://pandas.pydata.org/pandas-docs/stable/generated/pandas.date_range.html. Accessed on 07 Dec 2017

pandas: Time Series/Date functionality (2017c). http://pandas.pydata.org/pandas-docs/stable/timeseries.html#offset-aliases. Accessed on 07 Dec 2017

pandas: pandas.DatetimeIndex. o.J.-a. http://pandas.pydata.org/pandas-docs/stable/generated/pandas.DatetimeIndex.html . Accessed on 20 Dec 2017

PyGraphviz developer team: PyGraphviz (2016). https://pygraphviz.github.io/. Accessed on 03 Feb 2018

Python For Beginners: Python Docstrings. o.J. http://www.pythonforbeginners.com/basics/python-docstrings . Accessed on 11 Jan 2018

Python Software Foundation: PEP 318—Decorators for Functions and Methods (2003). https://www.python.org/dev/peps/pep-0318/. Accessed on 11 Jan 2018

Python Software Foundation: PythonDecorators (2016). https://wiki.python.org/moin/PythonDecorators. Accessed on 11 Jan 2018

Python Software Foundation: 2. Built-in Functions (2018a). https://docs.python.org/3/library/functions.html. Accessed on 11 Jan 2018

Python Software Foundation: 3. Data model (2018b). https://docs.python.org/3/reference/datamodel.html. Accessed on 12 Jan 2018

Python Software Foundation: 12.1. pickle—Python object serialization. (2018c).https://docs.python.org/3/library/pickle.html. Accessed on 20 Jan 2018

Python Software Foundation: msg 0.7.8: msg server backend. o.J. https://pypi.python.org/pypi/msg/0.7.8 . Accessed on 01 Mar 2018

Stack Overflow: What does the "at" (@) symbol do in Python? o.J. https://stackoverflow.com/questions/6392739/what-does-the-at-symbol-do-in-python . Accessed on 11 Jan 2018

Tamura, J., et al.: A new method of calculating in-plane irradiation by one-minute local solar irradiance. In: Proceedings of 3rd World Conference on Photovoltaic Energy Conversion, vol. 3, pp. 2265–2268. Japan (2003). http://ieeexplore.ieee.org/abstract/document/1305038/. Accessed on 14 Jan 2018

The Spyder Project Contributors: Spyder—Documentation. o.J. http://pythonhosted.org/spyder/ . Accessed on 22 Jan 2018

Chapter 5
Getting Started

You have now had an introduction to the Python programming language, to the OEMOF infrastructure and to creating an optimization model—now you can get started using the OEMOF system of programs. The first step is to install OEMOF on your computer. Once the installation has been completed, you can start creating your first model.

5.1 Installing OEMOF

How you install OEMOF depends on your operating system:

- Linux
- Windows
- Mac OSX

Detailed and up-to-date installation instructions for each operation system can be found at: http://oemof.readthedocs.io/en/stable/installation_and_setup.html. A tutorial for installing OEMOF for Windows can be found on youtube at https://www.youtube.com/watch?v=eFvoM36_szM: The following section describes the OEMOF installation for Windows 10. It is recommended to follow the instructions under the heading 'Using WinPython (community driven)' on the OEMOF hub site. Using Anaconda for Windows has caused a number of problems in the past.

There are four steps for installing OEMOF for Windows:

1. Installing Python 3 (WinPython)
2. Installing OEMOF
3. Selection and installation of a solver
4. Test run

Let us now go through those steps one by one:

© Springer International Publishing AG, part of Springer Nature 2019

J. Nagel, *Optimization of Energy Supply Systems*, Lecture Notes in Energy 69,

https://doi.org/10.1007/978-3-319-96355-6_5

Installing Python (WinPython)

The OEMOF tool is based on the Python programming language. This means that the computer OEMOF is installed on must have a Python 3 environment. Python 3 can be downloaded from the website of the OEMOF developer group (http://oemof. readthedocs.io/en/stable/installation_and_setup.html#windows).

To select the right Python version, e.g., WinPython, you must select the relevant operation system for your computer. The main difference is between 32-bit and 64-bit operating systems. How do I find out if my computer is 32- or 64-bit operated? For this we have a look at the Windows icon at the lower left-side corner of the screen (cf. Fig. 5.1).

If you click on this icon, the window displayed in Fig. 5.2 opens.

The information on the system type can be found under 'System' (cf. Fig. 5.3).

Once the system type has been obtained, you can download the relevant WinPython program from the WinPython website (cf. http://winpython.github.io/) (cf. Fig. 5.4). A link to this site can also be found on the OEMOF developers group site (cf. http://oemof.readthedocs.io/en/stable/installation_and_setup.html). As WinPython offers several different versions for 32/64-bit systems, it is a good idea to research online which version is currently recommended.

Next the directory where WinPython is to be stored must be selected. Of course, the program can be stored on any drive. But it is recommended to store the program on the main hard drive, such as C. Once the relevant version and a directory have been selected the program download can start. Fig. 5.5 shows the WinPython infrastructure after installation.

Windows icon, right mouse click

Fig. 5.1 Selecting the Windows icon

Selection button: System

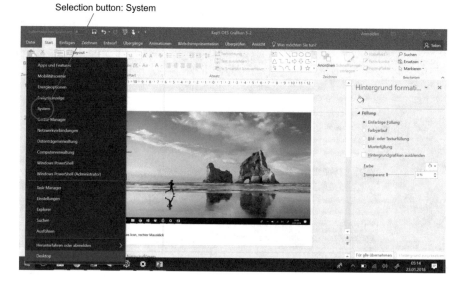

Fig. 5.2   Clicking on the Windows icon

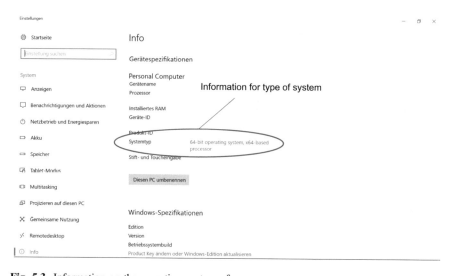

Fig. 5.3   Information on the operating system of your computer

After successfully installing WinPython step 2 is to install the OEMOF system of programs. For this the WinPython Command Prompt Console must be started (cf. Fig. 5.6).

Fig. 5.4 WinPython recent releases

Fig. 5.5 Infrastructure and directory for WinPython

Next you must give the command: *pip install oemof* in the prompt console (cf. Fig. 5.7). While OEMOF is still being installed, you can already start selecting the solver you want to use for solving the equations.

The website of the OEMOF developer group offers a download for a number of open source and commercial solvers (cf. Fig. 5.8). See also: http://oemof.readthedocs.io/en/stable/installation_and_setup.html#windows-solver-label.

WinPython Command Prompt Console

Fig. 5.6 Selection of the WinPython Command Prompt Console

Fig. 5.7 The command: *pip install oemof* on the WinPython Command Prompt Console

Fig. 5.8 Download options for solvers on the OEMOF developer site

There is a direct download link on the OEMOF website for solvers CBC and GLPK. It might be a good idea to select a second solver to verify your results. Another link refers to download sites for commercial solvers.

For selecting a solver, it is again important to know if you have a 32- or 64-bit system. The solver recommended by the OEMOF developer group is the CBC solver. The next section describes how to install this solver. After clicking on the link to the CBC solver download, you can download the file cbc-win64.zip for a 64-bit system and save it to your download directory. Before unzipping the file a directory for the solver must be chosen. Ideally, the directory should be a subdirectory of WinPython-64bit-3.6.3.0Qt5 (cf. Fig. 5.9).

Once the files have been extracted, the modeler has access to two key CBC files. First the environment variable must be set. For this the user must go into *System* to *Settings*. Using the field search settings, you can choose environment (cf. Fig. 5.10).

Fig. 5.9 Directory with files in the CBS directory

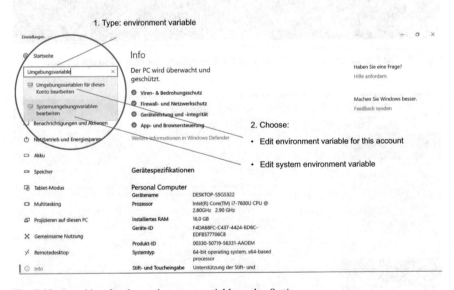

Fig. 5.10 Searching for the environment variable under *Settings*

The options available are:

– Change environment variables for this account
– Modify system variables

Choose the second option (cf. Fig. 5.11).

Click on this to get to the next screen, where you can change the environment variables. For this you have to select *Path* and then *Edit* (cf. Fig. 5.12).

The following screen allows you to create a path for the CBC solver (cf. Fig. 5.13). By clicking on *new*, a new field opens, into which you can write the path. This is completed by clicking on OK.

After the solver has been created, it should be tested, by entering the command *coemof_installation_test* in the WinPython Command Prompt Console. If this does not work, the solver can also be tested via Spyder. If the installation of the CBC solver was successful, the lower left-side window in Spyder will display the information that CBC has been installed and is running (cf. Fig. 5.14). The window also displays which solvers are not installed and not running.

Now that all installations have been completed, the system is ready for work. The users can now start to create their own models. A few more installations can

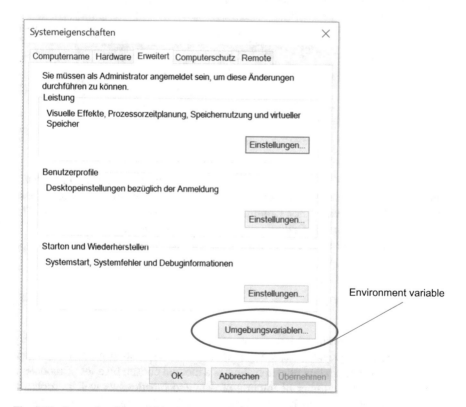

Fig. 5.11 Screenshot for modifying system variables

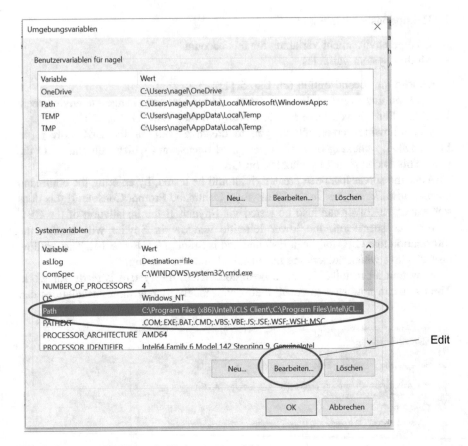

Fig. 5.12 Screen for editing the environment variables

help modelers in their work. Two more relevant installations are installing *feedinlib* and *demandlib*, as these two libraries are not part of the OEMOF package. These can be installed via WinPython Command Prompt Console by a pip command. The commands are: *pip install feedinlib* and *pip install demandlib*.

Once these two programs have been installed, the work can begin.

5.2 My First Model

We have already introduced key elements of creating your own model. One very common and easy system for creating a model is to use existing models and adapt them to your own modeling question. This is a method chosen by a lot of modelers, as this means that a number of imports of data and functions as well as tools for plotting and output of results are already available. Basing your model on an

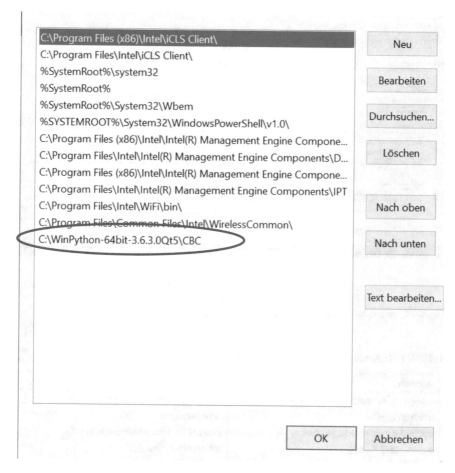

Fig. 5.13 Create a new path for the CBC solver

existing model can be a good way. OEMOF is flexible enough to allow both new models from scratch and reusing modified existing models.

It is recommended that future modelers have a look at completed models available from the OEMOF developer group at https://github.com/oemof. Modelers can put questions about the examples provided on the site or about their own models to OEMOF developers via https://oemof.org/contact/.

Before starting your model, you first have to decide if you are creating an investment or a dispatch model. The investment optimization model optimizes costs, in order to decide if plants are to be built or not. In contrast to investment models, dispatch models optimize the operation of plants. Table 5.1 compares and contrasts these two types of models.

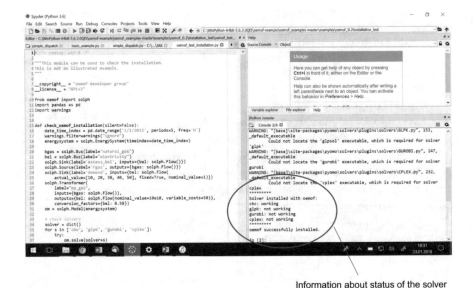

Information about status of the solver

Fig. 5.14 Status information for the solver in OEMOF

Table 5.1 Difference between parameters for a dispatch and an investment model

Dispatch	Invest
storage=components.GenericStorage(	storage=components.GenericStorage(
label='storage',	label='storage',
inputs={bel: Flow(variable_costs=0.1)},	inputs={bel: Flow(variable_costs=0.1)},
outputs={bel: Flow()},	outputs={bel: Flow()},
nominal_capacity=30,	—
capacity_loss=0.01,	capacity_loss=0.01,
initial_capacity=0,	initial_capacity=0,
nominal_input_capacity_ratio=1/6,	nominal_input_capacity_ratio=1/6,
nominal_output_capacity_ratio=1/6,	nominal_output_capacity_ratio=1/6,
inflow_conversion_factor=1,	inflow_conversion_factor=1,
outflow_conversion_factor=0.8)	outflow_conversion_factor=0.8,
—	investment=Investment(ep_costs=epc))

Chapter 1 already introduced the importance and key features of the objective function. To show how this can be implemented in OEMOF, find below a very simple example of an objective function (cf. Eq. 5.1):

$$y = \text{flow1} \cdot c1 + \text{flow2} \cdot c2 + \cdots \rightarrow \text{Min.} \qquad (5.1)$$

where

flow   model edges representing energy or cost flow
c weighting factors, such as emission costs

The parameters c ($c1$, $c2$,...) determine the weighting of a *flow* and thus determine what influence this flow will have on the overall result, which is to be as low as possible. Thus, if for a minimization model $c1$ is twice as high as $c2$ the solver will try to use flow2 instead of flow1. Parameters $c1$ and $c2$ can represent variable costs of specific emissions or other factors relevant for the optimization model. What value is returned for these parameters does not depend on the setup of the model but is determined by the parametrization. This can be illustrated by means of the following two equations. To show this, the same parameters $c1$ and $c2$ are given different values (cf. Eqs. 5.2 and 5.3).

$$\text{flow}_{\text{coal}} \cdot 6\frac{\text{ct}}{\text{kWh}} + \text{flow}_{\text{gas}} \cdot 9\frac{\text{ct}}{\text{kWh}} + \cdots \qquad (5.2)$$

or

$$\text{flow}_{\text{coal}} \cdot 99\frac{\text{gCO}_2}{\text{kWh}} + \text{flow}_{\text{gas}} \cdot 33\frac{\text{gCO}_2}{\text{kWh}} + \cdots \qquad (5.3)$$

The mathematical model is constructed in the same way for both equations. The cost and emissions flow can only be read after the values for the parameters have been entered. This also means that the same model can be used to create a climate change prevention study or to take an investment decision.

Calculating emissions, such as CO_2 emissions can also be a part of the constraints introduced into the model. Thus the CO_2 emissions could be introduced as an upper limit (cf. Eq. 5.4).

$$\sum \text{flow}_{\text{coal}} c_{\text{CO}_2} \leq \text{Max.} \qquad (5.4)$$

where

c_{CO_2} CO_2 emissions, e.g., in mg per kWh

The solver would then try to find a low-cost solution with low CO_2 emissions.

But the objective function can also be made more complex by adding annuity calculations into the investment decision. Such an objective function could then look similar like the one in Eq. 5.5.

$$y = \text{flow1} \cdot c1 + \text{flow2} \cdot c2 + \text{cap1} \cdot a1 + \text{cap2} \cdot a2 + \cdots \qquad (5.5)$$

where

cap   investment annuity
a installed capacity, e.g., of a wind power plant

There are two different approaches to creating models. A model can either be looking towards the future, i.e., when planning the extension or creation of an energy supply system. For a model of a future energy system both investment decision and operation strategy planning can be relevant (investment and/or dispatch models). But a model can also refer to an existing network of plants, for which the operation strategy is to be optimized (pure dispatch model).

Once these first key decisions have been taken, you can get started creating your first OEMOF-based model.

It is a good idea to first select an existing dispatch, investment or CHP operation model which is closest to the problem you are trying to model. This model can then be adapted to your own requirements step-by-step.

5.3   Defining Scenarios

Planning for the future is complex and will be influenced by several factors — social, political, technical, and economic. No one knows precisely what the future will look like. But in order to have some idea scenarios are developed. These scenarios can represent idealized visions of the future but also allow for planning operation strategies in order to make this vision become reality. Scenarios can be used to (Gausemeier 1996; von Reibnitz 1991; Götze 1993):

– Depict a vision of the future, by describing one possible future situation, and
– at the same time, describing a pathway that must be followed to make this future vision become reality.

Through the second aspect, describing developments that will further or hinder the planned development and describing possible development dynamics, a scenario differs from a pure utopian vision of the future (Gausemeier 1996; Götze 1993). Also, a scenario never describes all aspects of future life. A scenario is always focused on certain relevant aspects, such as the development of the oil price. For this reason a scenario can also be called a construction (Kosow 2008). This construction would then choose and combine some key factors within a defined timeframe. This means that the future is analyzed under a certain premise, e.g., a population increase of 20% within the next 20 years. This construction can only work if the knowledge base is good, which means that a lot of experience must be part of the knowledge base. As scenarios are hypothetical, their veracity can be expressed in probabilities. Scenarios are not the same as prognoses for the future. In contrast to prognoses, which are likely or possible future developments, scenarios are not based on the analysis of existing data. Prognoses can be produced on the basis of extrapolation of current and existing data and development tendencies. Examples would be weather prognoses and oil price development prognoses.

Scenarios serve to take a decision at the present time, in order to achieve future objectives. As part of the energy transition, energy stakeholders must decide which existing technologies they invest in and which new or potential new technologies merit further research. Another key aspect is which policies must be passed in order to achieve the climate protection targets. Taking such complex decisions normally requires more than one scenario. The conceptualization of the future will always be influenced by the modelers' views and assumptions on future development. The vision of the future is also based on assumptions of how the future will develop rooted in current and past development tendencies (Kosow 2008). Three separate points of view can be distinguished that developers might have on their own vision of the future (Grunwald 2002):

- The future can be calculated
  By means of the information on the past and the present, we can extrapolate what will happen in the future.
  For this approach, it makes sense to use past data in order to generate a vision of the future.
- The future evolves
  This approach is based on the assumption that the future evolves in a chaotic, aleatory and uncontrollable manner.
  For this approach, pathways and scenarios might be developed that are completely unconnected to past or present data and development tendencies. The influencing factors are chosen at random. A future scenario based on this approach will be rooted in chaos theory.
- The future can be shaped
  This approach assumes that we can influence the future by our actions. From this point of view, the future is neither chaotic nor predictable.
  If we can shape the future, then there is a certain probability that the chosen objective will be reached by means of the chosen pathway.

When we construct a scenario, we are influenced by these different points of view. This is certainly the case in our vision for our personal future, but it also has an effect on which method we choose when developing our scenario. Depending on the modelers' point of view, they will opt to focus on different influencing factors. Each of the selected influencing factors will have an impact on the scenario. It is very important to be aware of this strong bias in order to be susceptible to other viewpoints and other options for creating a scenario.

The selected influencing factors will guide our view on the future and are thus a key element of constructing a scenario. But these influencing factors are not inflexible, the further into the future we look the greater the amount of possible influencing factors and the number of possible development options. A space of future possibilities emerges. This space can be described by several distinct and separate scenarios. This means that we do not plot one future but a number of possible futures. These different scenarios are then compared, contrasted and evaluated as part of the decision-making process (Kosow 2008). The aim is to visualize possible future developments, in order to be able to play out the effects of different strategies and decisions on these possible futures.

Developing scenarios for the future is a key prerequisite for modeling the future. Especially in the energy sector, scenarios play a key role in deciding which technologies to invest in. Both Henning (2015) and Nagel (2017) use a development pathway scenario in order to show the potential advantages of creating a more flexible energy supply system. The chronological order for this development proceeds from the first step research via development and pilot projects to monitoring of the power plants and evaluation of their potential. Here (Henning 2015) assumes that the first step is a more flexible use of more conventional power plants, combined with a refurbishment and extension of the power grid. Power-to-heat plants play a major role in this scenario but also using synthetic fuels for producing heat and electricity towards the more distant end of the future development path.

Statements on future development are particularly difficult in the energy sector, as there are so many uncertainties to contend with. Prognoses in this sector often look 10, 20, or even 30 years into the future. As technological and social conditions might change drastically in these time periods and political decisions might be taken that are almost impossible to predict, it is even more important to create a future space of several possible scenarios.

In order to find a scenario to base your model on, you can either refer to published scenarios, such as the study of the German BDI on Germany's climate protection options (see Gerbert 2018), or you can develop your own scenarios.

Before creating a scenario, it is a good idea to set up a scenario workshop. The following section will introduce the key elements of a scenario workshop. The objectives of a scenario workshop are the following (Meyer 2009):

- Passing on information about future developments in the energy sector and discussing their likelihood. Pointing out risks and opportunities associated with them.
- Personal viewpoints and evaluations of possible obstacles for developing an energy supply model are to be identified and formulated by the participants of the workshop.
- Based on the expressed and assembled viewpoints, the views are pooled to create a joint scenario.

There are a number of different types of scenarios. The main types are:

- Explorative and normative scenarios:
  Explorative scenarios: the focus is on key factors and situational circumstances
  Normative scenarios: Take into account social viewpoints governing the desirability of future developments.
- Quantity or quality-oriented scenarios:
  Quantitative and qualitative scenarios.

– Timeframe-bound scenarios:
  e.g., medium-term or long-term scenarios.

The approach that is to govern the development of the scenario must be decided before the scenario workshop takes place. Other key requirements are the scenario question (such as 'What will the energy supply system look like in 2050?'), as well as setting objectives and the time limit and general framework, the so-called system boundaries. Once these points have been fixed, the scenario framework is defined.

The host's skill in making a good introduction to the scenario workshop by passing on key images and key points of the scenario to the participants will help to integrate everyone into the scenario process. The scenario boundaries are agreed together. These are assumptions that would be true for all scenarios developed in this group. Such assumptions might be the demographic development in Germany in the next decades. It is important to agree on such key assumptions at the beginning of the workshop.

The next step is to identify the key factors. Key factors are factors that will influence the system. They determine future development options and are thus central to the development of the scenario. Influencing factors can include legal requirements, but also the acceptance of renewable energies by the population and other aspects. At the beginning of the workshop the influencing factors should be identified by all participants together in the form of a brainstorming process. The chosen factors are then evaluated by the group as to their relevance and as to the uncertainty of their development.

At the end of the brainstorming and the evaluation process, the key factors will have been selected, these might include 'technology', 'CO_2 emissions cost', 'social acceptance' or 'climate change'. Each of these key factors must be further discussed as to its governing parameters. Ideally, more than one instance per factor should be identified. For example, when looking at the factor 'technology' instances could be different technologies, such as power-to-gas or energy storage. The key factor 'social acceptance' can be represented by selected specifications like 'ageing population' or 'desire for energy independence'. For each of the identified key factors possible future developments must be identified. This can be based on their importance or based on the uncertainty associated with them. By looking at possible developments for key factors, these factors are projected into the future. For cxample, two future projections for the key factor 'development of fuel prices' could be 'fuel prices remain constant' and 'fuel prices double'.

The scenarios thus also provide the relevant parameters for the mathematical optimization model. One possible parameter for the model could be the emission costs. Another parameter might be the energy technology. As part of the scenario choices, it would be decided which technologies will form part of the model and which not. This process of choosing parameters is described in a little more detail in the following section.

Developing a scenario is a complex process, as a number of the key factors carry a lot of uncertainty as to their development. As part of the scenario development it is possible to focus on specific aspects, i.e., on the positive or negative development

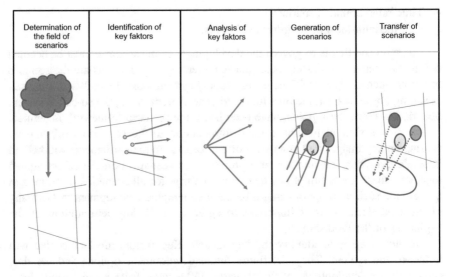

Determination of the field of scenarios	Identification of key faktors	Analysis of key faktors	Generation of scenarios	Transfer of scenarios

Fig. 5.15 The five phases of creating a scenario (based on Kosow 2008)

of one influence factor, such as the demographic development and then study its effect on energy demand and acceptance of wind power, and to combine these specific aspects and their interdependence into a detailed depiction of a possible future (Meyer 2009). This allows for a detailed and consistent description of a space of possible futures in the energy sector. By creating such a clear and detailed description, the factors for the model can be extracted from the scenario. The scenario process can be described through five phases (cf. Fig. 5.15).

In the first phase the question or problem at the heart of the scenario has to be agreed. What is the question that the scenarios are to be based on? What will be studied and what are the system boundaries? The next phase is to identify the key factors. These are to describe the scenario framework. Key factors affect the scenario framework or affect the outside frame of future events. For the model, they represent the variables, such as the switching on or off of wind power plants, or investment costs such as the development of the oil price. Specific events, such as the decision of Britain to leave the European Union (Brexit) can also be key factors in the process. Identifying the key factors can happen in different ways. This can be done by means of an analysis of data or by research of theoretical relations between developments or by using interactive workshops or surveys. The first phase is to analyze how the key factors will develop during the projected future and which instances they might develop. Phase four is generating the scenario. For this scenario key factors must be selected that are relevant and significant. These factors are pooled and scenarios based on them created. Sometimes more than one scenario has been developed or more than one scenario can be projected. But several parallel scenarios can be difficult to handle and cannot be analyzed in parallel from a cognitive point of view. From a practical point of view, it is recommended to not study more than four or five scenarios in one scenario framework (Kosow 2008). If too many scenarios are chosen, the modeling process can become difficult. If two few scenarios are chosen, some important aspects might be lost and some future possibilities discarded. The chosen scenarios must be clearly differentiated and delimited from each other so they allow for their own interpretation.

To choose a scenario there are several selection processes that have been tested and put into practice by several institutions (Greeuw 2000). These selection processes allow for four or more scenarios to be investigated simultaneously. The first process option is based on two logical assumptions (cf. Table 5.2). The first logical assumption allows modelers to formulate a statement about their future actions. For these actions, stakeholders start from the two extremes, from 'no measures are taken' to 'a lot of different measures are taken'. The second logical assumption is concerned with choosing the key influencing factors. Again, for choosing these, the two extremes as to their positive or negative development are studied.

Another scenario development process projects four dimensions with widely differing points of view onto future developments (cf. Fig. 5.16). This process almost automatically produces four separate scenarios.

Once these scenarios have been developed, the process is completed. For steps one to four, different methods can be implemented, some of which will be introduced in the following section. But the final process of the scenario selection is optional phase five, transferring the scenario. This phase can be used to analyze the effectiveness of scenarios. This phase also gives the opportunity to analyze the strategies for reaching your objectives. It is also possible to gauge the effect of scenarios onto the stakeholders. In our case the next step after creating the scenarios is the creation of mathematical models represented by an optimization function. Once this step has been completed, information about key factors and their effects on future developments can be obtained based on the modeled projections.

In the following section, we will introduce some scenario methodologies which can be used to complete phases one to four. The first of these methodologies are systematic and formalized scenario methodologies. These methodologies are only briefly introduced here and references are given for further reading. There are some commercial programs that can help you develop scenarios. As yet there are no open source programs available for this.

Systematic and formalized methodologies for scenario development

These methodologies entail defining and varying possible key factors and combining them with each other. This creates a type of scenario funnel (cf. Fig. 5.17) in which several scenarios can be created.

Table 5.2 Variation of point of view for creating measures and looking at key factors (Kosow 2008)

Scenario type	'Wait and see'	'Just do it'	'Doom monger'	'Carpe Diem'
Point of view	Measures		Development of scenario factors	
Logical basis	Few to none	Many	Positive	Negative

Fig. 5.16 Four points of view on the future (based on UNEP 2002)

Using this methodology, influencing factors are developed both based on quantitative and qualitative data. The data could be derived from a trend analysis, such as the development of energy prices or the basis of the analysis could be qualitative data, such as surveys on the acceptance of the energy transition in Germany. Based on this data, the key factors are identified by means of an influence analysis, which also assesses the interdependence of the different factors. The key question here is how these factors influence each other. This influence factor analysis can be carried out by means of a matrix, as shown in Table 5.3.

This matrix compares and contrasts factors and tests which interdependencies exist between paired factors (Wilms 2006). The following scale is recommended for assessing interdependencies (Blasche 2006):

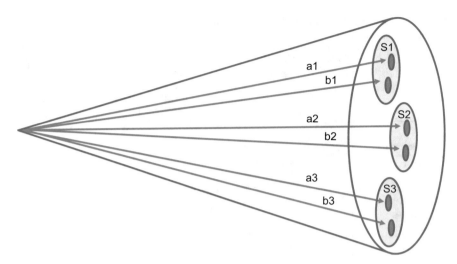

Fig. 5.17 Creation of a scenario funnel (based on Kosow 2008)

- 0 = no influence
- 1 = weak influence
- 2 = medium influence
- 3 = strong influence

By means of this scale, interdependencies can be quantified and sums for the lines and for the columns can be calculated. The line sum (LS) indicates how strong the influence of one factor on the other is. The column sum (CS) indicates how strongly one factor depends on another factor.

What is the outcome of this matrix? In order to understand the results, inter-dependencies between factors should be analyzed according to the following system (Kosow 2008):

- High LS and low CS
  Such factors are called active or impulsive factors. In Table 5.3 such factors are *electricity price* and *flexibilization*. These factors are not dependent on the scenario framework but themselves have a strong influence on the framework. If these factors can be changed, then modifying them will have a major influence on shaping the future. This means these factors will act as a lever or a switch for future developments.
- Low LS and high CS
  These are so-called reactive or passive factors. In Table 5.3 such factors are *emission costs* and *wind power acceptance*. These factors are more dependent on others and have little effect on the scenario framework. Such a factor is a useful tool for observing a situation.
- High LS and high CS

Table 5.3 Interdependency matrix of identified factors here shown with some example factors (based on Blasche 2006)

Effect on↗ from	Factor A emission costs	Factor B electricity price	Factor C wind power acceptance	Factor D average temperature of earth atmosphere	Factor E flexibilization of energy sector	Line sum (LS)
Factor A emission costs		0	0	2	0	2
Factor B electricity price	0		3	2	3	8
Factor C wind power acceptance	0	0		1	0	1
Factor D average temperature of earth atmosphere	3	1	3		0	7
Factor E flexibilization of energy sector	3	3	0	2		8
Column sum (CS)	6	4	6	7	3	

This is the case for critical and dynamic factors. A strong interdependence acts in both directions. Thus the scenario framework influences the factor and vice versa. This is due to the fact that these factors are strongly interconnected. These factors must therefore be observed very carefully. In our example in Table 5.3 this would be the case for the factor *average temperature of the earth atmosphere*.

– Low LS and low CS

These factors have a weak or buffering influence on the system. There is little influence of the scenario framework on them but they also do not influence the framework themselves. These factors are quite isolated and not interconnected to others. There is no example of such a factor in Table 5.3.

As only a few of the many potential influencing factors have been represented in Table 5.3 the interdependencies can only be shown as an example and do not allow for a detailed analysis.

But as can be seen, active, and critical factors will normally form the key factors chosen for a scenario. Weak and passive factors are not considered relevant for the scenario framework, as they are either considered to be stable or to only become relevant when they are interconnected to other factors, which they are currently not. The aim of such a matrix analysis is to identify a minimum of 10 to a maximum of 20 key factors (Kosow 2008).

The next phase is the analysis of the selected key factors. This analysis will be based on assumptions as to the future development of these factors. Assumptions can either be very conservative or 'creative', i.e., based on questions such as 'What is conceivable/inconceivable for this factor?' The scenarios will then be based on these assumptions. Let us look at the crude oil price for Brent as an example. On February 7, 2018 this was 65.50 $ (esyoil 2018). How might this price evolve? What are possible assumptions? Will the price continue to fall or start rising again? Will we have to assume a price of 100 $ for Brent crude oil in 2050 or is 200 $ more likely? What is the lower boundary? Is a further decrease of the crude oil price conceivable? Such as a price of 30.00 $? These are key questions that will always be subjective to a certain degree, i.e., dependent on our personal world view.

In order to develop consistent scenarios, these factors must be varied and combined with each other. One method for this is a consistency analysis (Heinecke 2006). This process is of particular relevance for gauging the interpretative value of scenarios (Gaßner 1992). But consistency analyses can also increase the modelers' faith in the validity of their scenario. As part of the consistency analysis different instances for each key factor are developed, for example for crude oil price the instance *crude oil price falls by 10%* and the instance *rises by 10%*. Probabilities are not part of this analysis, thus the question how likely a crude oil price fall by 10% is, is not relevant here. In order to test consistency, a factor is contrasted with other factors in its two instances, here referred to as (a) and (b) (cf. Table 5.4).

For each factor pair the consistency is evaluated. This means factor A in its instances (a) and (b) is compared to Factor B in its instances (a) and (b). To do this a scale from 1 to 5 is selected (Kosow 2008). The value 5 implies strong consistency, i.e., factors support each other; 4 implies a weak consistency, i.e., factors complement each other; 3 represents a neutral or independent relationship between factors; 2 implies a weak inconsistency, this means that factors are not fully compatible and 1 indicates that factors are completely incompatible with each other. Once the consistency of each factor in its two instances has been assessed, the factors can be combined and pooled. The result is a set of pooled factors with their instances, which can each represent a basic scenario. In order not to have to deal with too many different scenarios, those factors and instances with high consistency will be selected. This reduces the number of factor sets.

Table 5.4 Example of a consistency matrix (based on Gausemeier 1996)

How does the instance of the line factor influence the instance of the column factor?		Factor A		Factor B		Factor C		Factor D	
		Instance A-a)	Instance A-b)	Instance B-a)	Instance B-b)	Instance C-a)	Instance C-b)	Instance D-a)	Instance D-b)
Factor A	Instance A-a)								
	Instance A-b)								
Factor B	Instance B-a)	3	5						
	Instance B-b)	4	3						
Factor C	Instance C-a)	4	3	5	2				
	Instance C-b)	3	4	5	2				
Factor D	Instance D-a)	4	1	1	4	3	5		
	Instance D-b)	1	3	5	2	1	3		

Trend extrapolation methodology for constructing scenarios

At the beginning of this chapter it was explained that scenarios are not prognoses that are based on trend extrapolation. Nevertheless, some key quantitative factors can be varied by means of trend extrapolation or trend analysis. For this, past and present trends are projected into the future. A trend is a long-term development of a parameter. Figure 5.18 shows such a trend extrapolation for the average temperature for the earth atmosphere.

A trend extrapolation means that development tendencies that are in evidence in statistical data are plotted into the future, as shown in Fig. 5.18. In contrast to trend extrapolation, trend analysis is concerned with analyzing, quantifying and searching for causes of long-term development tendencies for a factor. Thus a trend analysis is more complex and more comprehensive than a trend extrapolation.

The methodology of trend extrapolation is the gathering of long-term data and information. This trend is then projected into the future by means of statistical methods, such as time series analysis.

Trend extrapolation does not create a space of possible futures but instead describes one particular future, i.e., one particular development, which is considered probable based on statistical assumptions (Kosow 2008). This means that this method is not conducive to producing more than one possible future scenario, but it

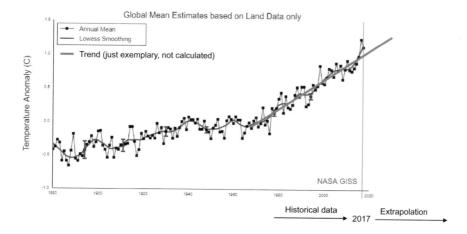

Fig. 5.18 Global annual mean surface air temperature change (based on NASA 2018)

can be useful for creating a reference scenario. Other scenarios can then be compared to this reference scenario. When making use of extrapolation, one must be aware of the fact that extrapolation suggests a certainty of this development which simply reflects a certain statistical probability. Major discrepancies can still occur. For this reason, this scenario development methodology cannot be recommended for the energy sector.

But an improved development of this method, the so-called trend-impact analysis can be used to further describe some key factors. By means of trend-impact analysis, factors can be assigned different instances. To do this, first trend extrapolation is used to plot a 'highly probable' development (Gordon 1994). As a next step, the development curve is modified by projecting future results based on expert interviews and then calculating their result on the 'highly probable' development. This additional data then provide for trend development alternatives as shown in Fig. 5.19.

This method thus also provides for alternatives by means of which additional scenarios can be generated.

Developing scenarios based on creative and narrative methodologies

This methodology strongly focuses on the scenario development process. Once key factors have been identified, the scenario development itself can be based on a permutation methodology (Steinmüller 2002). For this, some selected key factor instances are combined with each other. It is important to limit the overall number of key factors used, in order to be able to keep track of the process. A matrix is developed in which the instances are assessed according to certain criteria, such as high or low difference between instances as well as excess or deficit effect (cf. Fig. 5.20).

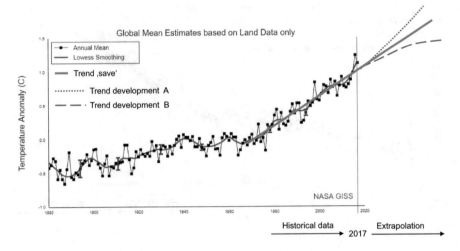

Fig. 5.19 Example of a trend-impact analysis (based on NASA 2018)

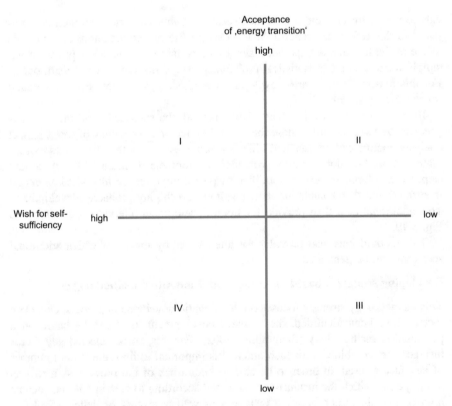

Fig. 5.20 Matrix creation by means of the permutation methodology (based on Steinmüller 2002)

The matrix then allows to create a description of the future for each of the quadrants. Based on these descriptions the key factor instances can be pooled into scenarios again. But the number of scenarios selected should still be limited to three to four, as this is enough to describe the space of possible futures. Once these basic scenarios are selected, they must be described in more detail.

The scenarios are the key for the creation of a model, as the scenarios provide projected values for the parameters that are to be entered into the model. Once the values for the parameters have been selected, the optimization calculations can start.

References

Blasche, U.G.: Die Szenariotechnik als Modell für komplexe Probleme: Mit Unsicherheiten leben lernen. In: Falko E.P. Wilms (ed.) Szenariotechnik: Vom Umgang mit der Zukunft, pp. 61–92. Haupt Verlag, Bern (2006)

esyoil GmbH: Ölpreis aktuell: Ölpreis News & Ölpreisentwicklung (2018). https://www.esyoil.com/%C3%B6lpreis. Accessed 08 Feb 2018

Gausemeier, J. et al.: Szenario-Management: Planen und Führen nach Szenarien. 2., bearbeitete Auflage, Hanser Verlag, München, 1996

Gaßner, R.: Plädoyer für mehr Science Fiction in der Zukunftsforschung. In: Burmeister, K., et al. (eds.) Streifzüge ins Übermorgen, pp. 223–232. Beltz Verlag, Weinheim (1992)

Gerbert, P. et al.: Klimapfade für Deutschland. Studie im Auftrag des Bundesverbandes der Deutschen Industrie (BDI), Berlin, 2018. https://bdi.eu/publikation/news/klimapfade-fuer-deutschland/, Accessed on 24 Jan 2018

Gordon, T. J.: Trend Impact Analysis. In: Glenn, J. C. et al. (Ed.): Futures Research Methodology Version 3.0. CD-ROM. The Millennium Project, 1994. http://107.22.164.43/millennium/FRM-V3.html, Accessed on 06 June 2018

Götze, U.: Szenario-Technik in der strategischen Unternehmensplanung. 2., aktualisierte Auflage. Wiesbaden (1993)

Greeuw, S. C.H. et al. (Authors); International Centre for Integretive Studies (ICIS): Cloudy Crystal Balls : An assessment of recent European and global Scenario studies and Models. Experts' corner report. Prospects and scenarios No 4. Environmental issues series no 17, European Environment Agency, Copenhagen, 2000

Grunwald, A.: Technikfolgenabschätzung: Eine Einführung. Edition Sigma, Berlin (2002)

Heinecke, A.: Die Anwendung induktiver Verfahren in der Szenario-Technik. In: Falko E. P. Wilms (ed.) Szenariotechnik. Vom Umgang mit der Zukunft, pp. 183–213. Haupt Verlag, Bern (2006)

Henning, H.-M. et al.: Phasen der Transformation des Energiesystems. In: Energiewirtschaftliche Tagesfragen, 65. Jahrgang, Heft 1/2; pp. 10-13. EW Medien und Kongresse GmbH, Essen, 2015. http://www.et-energie-online.de/AktuellesHeft/Topthema/tabid/70/NewsId/1230/Phasen-der-Transformation-des-Energiesystems.aspx, Accessed on 28 Feb 2018

Kosow, H. et al.: Methoden der Zukunfts-und Szenarienanalyse: Überblick, Bewertung und Auswahlkriterien. IZT WerkstattBericht Nr. 103. IZT – Institut für Zukunftsstudien und Technologiebewertung, Berlin, 2008. https://www.izt.de/fileadmin/publikationen/IZT_WB103.pdf, Accessed on 24 Jan 2018

Meyer, R. et al.: Diskursprojekt "Szenario Workshops: Zukünfte der Grünen Gentechnik": Leitfaden "Szenario Workshop". BMBF Förderkennzeichen: 01GP0774, Karlsruhe et al., 2009. http://www.itas.kit.edu/pub/v/2009/meua09g.pdf, Accessed on 24 Jan 2018

Nagel, J.: Energie- und Ressourceninnovation: Wegweiser zur Gestaltung der Energiewende. Hanser Verlag, München (2017)

NASA National Aeronautics and Space Administration: GISS surface temperature analysis (2018). https://data.giss.nasa.gov/gistemp/graphs/. Accessed 09 Feb 2018

Steinmüller, K.: Workshop Zukunftsforschung. Teil 2: Szenarien: Grundlagen und Anwendungen. Z_punkt GmbH, Essen (2002)

UNEP—United Nations Environment Programme: GEO 3 Global environment outlook 3. In: Past, Present and Future Perspectives, pp. 2002–2032. UNEP, Nairobi (2002)

von Reibnitz, U.: Szenario-Technik: Instrumente für die unternehmerische und persönliche Erfolgsplanung. Wiesbaden (1991)

Wilms, F.E.: Szenariotechnik: Vom Umgang mit der Zukunft. Haupt Verlag, Bern (2006)

Chapter 6
Modeling Optimization Problems in the Energy Sector with OEMOF

This chapter will introduce some examples of modeling with OEMOF. Based on OEMOF Release v0.1.x a study of the B.A.U.M. Consult Group is introduced which makes use of the class *Transformer* defined as one of two types, either one input and several outputs or several inputs and one output. The example shown here is a specific model developed by a business consultancy company. These OEMOF examples have been chosen as indicative of modeling options. Unfortunately, there are not yet any published studies available based on Release v0.2.0.

6.1 Study 'Electrification of Agricultural Machinery by Means of Solar Energy'

This study was carried out as part of the joint projects SESAM (FKZ 01ME12124), GridCon (FKZ 01ME14004C) (www.gridcon-project.de) and GridCon2 (FKZ 01ME17007C on behalf of the German Federal Ministry for Economic Affairs and Energy (Stöhr 2018a). A research group composed of the manufacturer John Deere, the university TU Kaiserslautern and B.A.U.M. jointly carried out this study on developing and setting up a connection to the electric grid for agricultural machinery (projekt-gridcon 2018). The objective of the project is to protect the climate by using batteries and solar panels for creating agricultural machinery running on electricity rather than diesel or petrol (baumgroup 2018). As part of the study the use of fully electric agricultural machinery in the context of using solar energy and the effect on the local grid is investigated. Agricultural machines can either be electrified by using electric batteries or by connecting agricultural machines to the local electricity grid by means of a cable. If more agricultural machinery is electrified, this has an effect on the local electricity grid. For this reason it was decided to include the effect of electrifying agricultural machinery onto local country electricity grids in the study. The future smart grid infrastructure

© Springer International Publishing AG, part of Springer Nature 2019
J. Nagel, *Optimization of Energy Supply Systems*, Lecture Notes in Energy 69,
https://doi.org/10.1007/978-3-319-96355-6_6

would allow to balance the electricity demand from renewable energies for electrifying agricultural machinery and local demand for electricity for heating buildings or running households or manufacturing, as smart grids would be able to balance demand by switching between supporting local demand and storing to stationary storage units.

The B.A.U.M. Consult GmbH based in Munich, Germany, developed an optimization model in OEMOF, in order to determine the most economically viable and the most ecologically advantageous operation of an electric agricultural machine. For this study two modules were created in OEMOF. The first module was used to implement the economic viability approach (economics_BAUM) (Stöhr 2018b). The second module was used to create an application for the optimization model (GridCon_storage) (Stöhr 2018b). Both programs are based on Release v0.1. The application optimizes costs for the extension of the local electricity grid and for adding stationary electricity storage.

The future vision for a fully electric agricultural machine could be represented by an autonomous vehicle connected by cable to an electricity socket at the end of the field, as would be the case for a fully automatic lawnmower or vacuum cleaner. The other option available to date would be using energy storage. By means of a charging point where empty batteries could be exchanged against recharged batteries a smooth running of the machines could be ensured.

6.1.1 Calculating Investment in OEMOF

The OEMOF program system provides the module *economics* under its *tools* heading (GitHub 2018). The example given here is assessing the economic viability of an investment by means of annuities. The viability is calculated and returned by means of an optimization function.

Annuity is one of several possible methods for assessing the economic viability of an investment. This is a type of dynamic investment assessment. When assessing investment dynamically, the interest rate payable on capital is also taken into account (Konstantin 2017). These methods are used particularly when looking at long-term investment, which is very often the case with investments in the energy sector. One particularity of the annuity calculation method is the fact that the asset lifetime is not taken into account (Konstantin 2017). This is significant, as the lifetime for power plants and other assets can differ considerably in the energy sector. For example, batteries have a much lower lifetime than wind turbines. By using annuities as reference, these two assets can both be included in the same mathematical equation.

The process for creating an investment optimization function in OEMOF is the following. In OEMOF investment is normally seen as discontinuous investment (Stöhr 2018b). This type of financing must be set against other regular cost flows. Since these two cost flows differ in structure, they cannot be calculated in one financial equation. Therefore, investment is calculated as an annuity payable in recurring annual increments. This is the annuity that can be calculated as shown in Eq. 6.1 (Stöhr 2018b):

$$I_0 + \sum_{i=1}^{n} I_i \left(\frac{1}{1+z}\right)^i = A \sum_{i=1}^{n} \left(\frac{1}{1+z}\right)^i \tag{6.1}$$

where

I_0 initial investment
i investment per year i
n period of n years
z interest rate for calculations
A annuity

Equation 6.1 can be interpreted in the following way: The discontinuous investments are listed on the left side of the equation. These investments are calculated as a once a year regular payment (Stöhr 2018b). This is what is described as the annuity. For the further transpositions of the equations the factor $\left(\frac{1}{1+z}\right)$ is subsumed under q (cf. Eq. 6.2):

$$q = \frac{1}{1+z} \tag{6.2}$$

The formula for sums for geometric series can be applied here (cf. Eq. 6.3) (Konstantin 2017):

$$\sum_{i=1}^{n} q^i = \frac{q(1-q^n)}{1-q} \tag{6.3}$$

In OEMOF investment calculations are normally carried out in such a way that only one first investment is paid during the time period that is analyzed. No further investment takes place. For this special case Eq. 6.1 can be modified in the form of Eq. 6.3 to become Eq. 6.4 (Stöhr 2018b):

$$A = I_0 \frac{1-q}{q(1-q^n)} = I_0 z \frac{(1+z)^n}{(1+z)^{n-1}} \tag{6.4}$$

For optimization calculations based on investment OEMOF regularly uses Eq. 6.4 in the module *economics*. The program then returns the variable *equivalent periodical (annual) costs (epc)* for an investment. The equivalent periodical costs (epc) are the factor that is to be optimized in the objective function of the optimization model.

In the module *economics* this is implemented in the source code as shown in the excerpt (GitHub 2018):

```
# -*- coding: utf-8 -*-
```

```
""" Module to collect useful functions for economic calculation.
This file is part of project oemof (github.com/oemof/oemof). It's copyrighted
by the contributors recorded in the version control history of the file,
available from its original location oemof/oemof/tools/economics.py
SPDX-License-Identifier: GPL-3.0-or-later
"""
```

```python
def annuity(capex, n, wacc):
    """ Calculate the annuity.
    annuity = capex * (wacc * (1 + wacc) ** n) / ((1 + wacc) ** n - 1)
    Parameters
    ----------
    capex : float
        Capital expenditure (NPV of investment)
    n : int
        Number of years that the investment is used (economic lifetime)
    wacc : float
        Weighted average cost of capital
    Returns
    -------
    float : annuity
    """
    return capex * (wacc * (1 + wacc) ** n) / ((1 + wacc) ** n - 1)
```

6.1.2   Finance-Mathematical Extension of the Annuity Method in the Modul Economics_BAUM

In the study conducted by the B.A:U.M. Consult as part of the GridCON project, the equivalent periodical costs (epc) of an investment are carried out in the module *economics_BAUM* as in the OEMOF module. The variable *epc* of the module contains all fixed annual costs for running a plant.

The approach described in the module *economics* in OEMOF is insufficient to describe the system interdependencies studied here, which is why the new module *economics_BAUM* was developed. As this study looks at lithium-ion battery storage, it is important to refer to the very short lifetime of the batteries compared to the longer lifetime of a grid and also to refer to the decreasing investment costs for these batteries. The short lifetime of these batteries is the reason why more than one investment for the purchase of new batteries will have to be entered into the calculation in the period studied. Furthermore, the decreasing costs for batteries had to be reflected by means of a function with an exponential decrease. The period studied by B.A.U.M. Consult in financial terms was the lifetime of the hardware of an electric grid (Stöhr 2018b).

In order to be able to take all these additional aspects into account, Eq. 6.1 was extended in the following way (cf. Eq. 6.5) (Stöhr 2018b):

$$I_0 + \sum_{j=1}^{m-1} I_0(1 - cd)^{ju}q^{ju} = A\sum_{i=1}^{n} q^i \tag{6.5}$$

where

i continuous counter of the year
j continuous counter of follow-up investment
m number of follow-up investment
u hardware lifetime of investment
n $n = m \cdot u$ — i.e., the hardware lifetime end of the last follow-up investment corresponds to the end of the period studied
I_0 initial investment
cd $cd = \frac{I_i - I_{i+1}}{I_i}$ — annual cost decrease

If the first follow-on battery is purchased after u years, the nominal costs for this can be calculated by means of Eq. 6.6:

$$I_0^u = I_0(1 - cd)^u \tag{6.6}$$

If we now look at the jnth follow-on investment, that will be required after jxu years, the cost calculation is modified as shown in Eq. 6.7:

$$I_0^{ju} = I_0(1 - cd)^{ju} \tag{6.7}$$

By means of the factor q^{ju} nominal values are transformed into calculable values.

Now the geometric series from Eq. 6.3 is applied to Eq. 6.5. This produces Eq. 6.8 (Stöhr 2018b):

$$I_0\left(1 + ((1 - cd)q)^u \frac{1 - ((1 - cd)q)^{(m-1)u}}{1 - ((1 - cd)q)^u}\right) = A\frac{q(1 - q^n)}{1 - q} \tag{6.8}$$

If you solve this equation for annuity A the result is Eq. 6.9:

$$A = I_0 \frac{1 - q}{q(1 - q^n)} \cdot \frac{1 - ((1 - cd)q)^{mu}}{1 - ((1 - cd)q)^u} \tag{6.9}$$

Now, instead of factor q the original term $\frac{1}{1+z}$ is used again. And factor $m \cdot u$ is replaced by n t. This leads to Eq. 6.10:

$$A = I_0 z \frac{(1+z)^n}{(1+z)^n - 1} \cdot \frac{1 - \left(\frac{1-cd}{1+z}\right)^n}{1 - \left(\frac{1-cd}{1+z}\right)^u} \tag{6.10}$$

In the module *economics_BAUM* the variables are renamed in accordance with their financial significance:

I_0 capex — investment expenditure
z wacc — weighted average cost of capital
cd   cost_decrease

The term a_n in Eq. 6.11 represents the annuity factor:

$$a_n = z \frac{(1+z)^n}{(1+z)^n - 1} \tag{6.11}$$

When taking into account the renaming of factors, Eq. 6.11 turns into Eq. 6.12:

$$a_n = \text{wacc} \frac{(1+\text{wacc})^n}{(1+\text{wacc})^n - 1} \tag{6.12}$$

And consequently, Eq. 6.10 corresponds to Eq. 6.13:

$$A = \text{capex} \cdot a_n \cdot \frac{1 - \left(\frac{1-cd}{1+\text{wacc}}\right)^n}{1 - \left(\frac{1-cd}{1+\text{wacc}}\right)^u} \tag{6.13}$$

If required, the factor oc can be used to represent all fixed annual operational costs to Eq. 6.13. The factor oc then becomes a further variable for the calculation.

The value obtained from Eq. 6.13 is called the equivalent periodical (annual) costs (epc) of an investment. This is the value that will be returned after conducting the optimization calculation in OEMOF, complemented, if required, by the periodic fixed operational costs oc. The term capex reflects the modification of the annuity for repeated investment costs. In this calculation, the follow-up investment takes place in fixed intervals u and with continually decreasing associated costs. If the periods are annual, then the epc corresponds directly to the annuity.

The 'new' energy sector encompassing renewable energies at a greater level allows the consumers to become energy producers themselves and to feed the generated electricity back to the grid. As this means that consumers can also be producers, these are sometimes called *prosumers*. This becomes relevant for modeling the energy sector, because these prosumers now receive a fixed income via selling electricity. This fixed income could previously not be represented in the

OEMOF modeling system. By means of the *economics_BAUM* module, this can now be integrated into modeling applications.

As part of the GridCON study this module was used to calculate fixed income payments to prosumers for primary energy capacity (PRL) by means of a stationary energy storage unit. In order to be able to do this, the new variable *oc* was extended to represent the difference between fixed annual costs and fixed annual income from PRL capacity (Stöhr 2018b). This variable now represents the fixed annual net electricity costs after deduction of projected fixed payments for PRL. Thus the variable now no longer simply reflects fixed annual operational costs.

6.1.3 The Source Code of the Module Economics_BAUM

The following section introduces the source code of the module *economics_BAUM*. Additional information on each of the programming steps is provided directly in the source code in the # sections (Stöhr 2018b):

```
# -*- coding: utf-8 -*-
""" " "
Module to collect useful functions for economic calculations.
""" " "


def epc(capex, n, u, wacc, cost_decrease = 0, oc = 0):

    """ " "
    function represents equivalent periodical costs of an economic
    activity

    parameters
    ----------
    capex: float
        capital expenditure for first investment per functional unit
        (e.g. 1 kWh or 1 kW)
    n : int
        number of years investigated; might be an integer multiple
        of technical
        lifetime (u) in case of repeated investments; in case of a single
        investment, n must equal u
    u : int
        number of years that a single investment is used, i.e. the
        technical
        lifetime of a single investment
```

```
wacc : float
    weighted average cost of capital
cost_decrease : float
    annual rate of cost decrease for repeated investments
    takes the value "0" if not set otherwise, that is in case of a single
    investment or in case of no cost decrease for repeated investments
oc : float
    fixed annual operational costs per functional unit
    (e.g. 1 kWh or 1 kW)
    takes the value "0" if not set otherwise

" " "

annuity_factor = (wacc*(1+wacc)**n)/((1+wacc)**n-1)

#  investment  (annuity)  and  the  costs  of  investment  in  case  of  a
#  single initial investment

return (annuity_factor*capex*((1-((1-cost_decrease)/(1+wacc))**n)
                        /(1-((1-cost_decrease)/(1+wacc))
                        **u)))+oc

# the value returned are the equivalent periodical (annual) costs of an
# investment (annuity) plus the periodical fixed operational costs (oc);
# the expression behind "capex" reflects the modification of the annuity
# in case of repeated investments at fixed intervals u with decreasing costs;
```

6.1.4   The Application GridCon_Storage

The next section introduces the application *GridCon_storage* for optimizing the costs for expanding the electric grid to support electricity demand and to include stationary storage units. The source code here corresponds to the general rules for creating source code in OEMOF and also contains detailed documentation.

```
###################################################################
# IMPORTS
###################################################################

# outputlib

from oemof import outputlib

# default logger of oemof

from oemof.tools import logger
from oemof.tools import helpers
from oemof.tools import economics_BAUM

    # economics is a tool to calculate the equivalent periodical cost (epc)
    # of an investment;
    # it has been modified by B.A.U.M. Consult GmbH within the frame of the
    # GridCon project (www.gridcon-project.de) and the modified version
    # has been called "economics_BAUM";
    # it allows now calculting epc of a series of investments with a
    # defined cost-decrease rate;
    # it also allows taking into account fixed periodical costs such as
    # staff cost and offset fixed periodical income;

from pyomo import environ
import oemof.solph as solph

# import oemof base classes to create energy system objects

import logging
import os
import pandas as pd

try:
    import matplotlib.pyplot as plt
except ImportError:
    plt = None

# import load and generation data from csv-file and define timesteps

def optimise_storage_size(filename="GridCon1_Profile.csv",
                          solver='cbc', debug=True, number_timesteps=
                          (96*366), tee_switch=True):

    # the file "GridCon1_Profile" contains the normalised profile for the
```

```
    # agricultural base load profile L2, a synthetic electrified
    # agricultural machine load profile, and the PV generation profile ES0;
    # number_timesteps: one timestep has a duration of 15 minutes,
    # hence, 96 is the number of timesteps per day;
    # 366 is the number of days in a leap year, chosen here because
    # standard load profils of 2016 are used;
    # the total number of timesteps is therefore 96*366 = 35136;

# initialise energysystem, date, time increment

    logging.info('Initialise the Energysystem')

    date_time_index = pd.date_range('1/1/2016', periods=number_time
    steps,

freq='15min')
    energysystem = solph.EnergySystem(timeindex=date_time_index)

    time_step = 0.25

    # a 15 minutes time step equals 0.25 hours;

# read data file

    full_filename = os.path.join(os.path.dirname(__file__), filename)
    data = pd.read_csv(full_filename, sep=",")

#####################################################################
# DEFINITION OF TO-BE-OPTIMISED STRUCTURES
#####################################################################

# definition of the investigated (financial) period for which the
# optimisation is performed;
# needs to be the same for all objects whose costs are taken into account;

    n = 50

    # financial period in years for which equivalent periodical costs of
    # different options are compared;

# definition of the specific investment costs of those objects whose size
# is optimised;
# here, the electric grid connection and the electrical storage;
# the electric grid connection considered here comprises the local mv-lv
# transformer plus the respective share of the entire up-stream grid;
```

```
# as a consequence of oemof allowing to handle only positive flow values,
# the grid connection needs to be modelled twice: a "collecting half" for
# the electric power flow from the local lv-grid to a far point in the
# up-stream grid (it collects electricty generated in areas where
# generation exceeds the demand at a given moment), and a "supplying half"
# for the inverse flow from that far point to the local grid (it supplies
# areas where the demand exceeds the generation at a given moment);

    invest_grid = 500

    # assumed specific investment costs of the electric transformer linking
    # the low voltage and the medium voltage grid including respective
    # share of up-stream grid costs in €/kW;
    # the value is taken from a real price (about 200000 €) paid by an
    # investor for grid connection of about 400 kW active power provision
    # capacity (at the low-voltage side of the transformer) set up for a
    # new large load in a rural area; this amount contains essentially
    # upstream grid costs;
    # source: oral communication from a private investor;

    invest_el_lv_1_storage = 300

    # assumed specific investment costs of electric energy storage system
    # in €/kWh;
    # figure reflects roughly specific investment costs of lithium-ion
    # batteries;
    # source: Sterner/Stadler, Energiespeicher, p. 600
    # (indicates 170 - 600 €/kWh)

# definition of parameters entering in the calculation of the equivalent
# periodical costs (epc) of the electric transformer and the up-stream
# grid;

    wacc = 0.05

    # assumed weighted average cost of capital;

    u_grid = 50

    # assumed technical lifetime of electric transformer and up-stream
    # grid;

    cost_decrease_grid = 0
```

```
# indicates the relative annual decrease of investment costs;
# allows calculating the cost of a second or any further investment
# a certain number of years after the first one;
# here, only one investment in the electric transformer and up-stream
# grid is considered to be made here within the financial period;
# hence, there is no cost decrease and the variable takes the value
# zero;


oc_rate_grid = 0.02


# percentage of initial investment costs assumed for calculation of
# specific annual fixed operational costs of electric transformer and
# up-stream grid;


oc_grid = oc_rate_grid * invest_grid


# specific annual fixed operational costs of electric transformer and
# up-stream grid in €/kW of active power provision capacity;

# calculation of specific equivalent periodical costs (epc) i.e. the annual
# costs equivalent to the investment costs (annuitiy) plus the fixed
# operational costs of the electric transformer and the up-stream grid per
# kW of active power provision capacity;


sepc_grid = economics_BAUM.epc(invest_grid, n, u_grid, wacc,
                               cost_decrease_grid, oc_grid)


# specific equivalent periodical costs of transformer and of up-stream
# grid in €/kW;

# definition of parameters entering in the calculation of the equivalent
# periodical costs of the electric energy storage system;


u_el_lv_1_storage = 5


# assumed technical lifetime of electric energy storage system


cost_decrease_el_lv_1_storage = 0.1


# assumed annual cost decrease rate for newly installed electric energy
# storage systems; reflects roughly learning curve for lithium-ion
# battery storage systems in 2010-2016;


oc_rate_el_lv_1_storage = 0.02
```

```
# percentage of specific initial investment costs assumed for
# calculation of specific annual fixed operational costs of electric
# storage system;

oc_el_lv_1_storage = oc_rate_el_lv_1_storage * invest_el_lv_1_storage

# specific annual fixed operational costs of electric storage system in
# €/kWh;
```

```
# calculation of specific equivalent periodical costs (sepc in €/kWh/year)
# i.e. the specific annual costs equivalent to the investment costs
# (annuitiy) plus the fixed operational costs of the electric energy
# storage system;

sepc_el_lv_1_storage = economics_BAUM.epc(invest_el_lv_1_storage, n,
                                          u_el_lv_1_storage, wacc,
                                          cost_decrease_el_lv_1_storage,
                                          oc_el_lv_1_storage)

kS_el = sepc_el_lv_1_storage

# equivalent specific annual costs of electric energy storage system in
# €/kWh;
# refers to nominal storage capacity;
```

```
# calculation of income from provision of primary balancing power by
# electric energy storage system; income is substracted from equivalent
# periodical costs;

prl_on = 0

# if primary balancing power is planned to be provided by the energy
# storage, set value "1", otherwise "0";

prl_weeks = 13

# number of entire weeks for which primary balancing power is planned
# to be provided

prl_income = 2.4 * prl_on * prl_weeks

# corresponds to specific annual income per kWh of nominal electric
# energy storage capacity, i.e. expressed in €/kWh, generated by
```

```
# provision of primary balancing power in Germany at a remuneration of
# 3000 €/week by an energy storage with a charge/ discharge rate of at
# least 1 MW per MWh of storage capacity operated between 10% and 90%
# of its nominal capacity;

sepc_el_lv_1_storage = sepc_el_lv_1_storage – prl_income

kS_el_netto = sepc_el_lv_1_storage

# net specific equivalent periodical costs of electric energy storage
# system taking into account income generated from provision of primary
# balancing power;
```

```
####################################################################
# CREATION OF OEMOF STRUCTURE
####################################################################

    logging.info('Constructing GridCon energy system structure')
```

```
####################################################################
# CREATION OF BUSES REPRESENTING ENERGY DISTRIBUTION
####################################################################

    b_el_mv = solph.Bus(label="b_el_mv")

    # creates medium voltage electric grid

    b_el_lv = solph.Bus(label="b_el_lv")

    # creates low voltage electric grid
```

```
####################################################################
# CREATION OF SOURCE OBJECTS
####################################################################

    solph.Source(label='mv_source', outputs={b_el_mv: solph.Flow()})

    # represents aggregated electric generators at a far point in the
    # up-stream grid; here, no limit is considered for this source;

    solph.Source(label='el_lv_7_pv', outputs={b_el_lv:
        solph.Flow(actual_value=data['pv'], nominal_value = 1,
                    fixed=True)})

    # represents aggregated pv power plants in investigated area which are
```

```
# looked at as a single source of energy;
# "outputs={b_el_lv: ...}" defines that this source is connected to the
# low voltage grid;
# "solph.Flow ..." defines properties of this connection: actual_value
# get the pv generation data for all time intervals from csv-file;
# "nominal_value = 1" signifies that pv generation data do not need
# further processing, they are already absolute figures in kW;
# "fixed=True" signifies that these data are not modified by the
# solver;

  solph.Source(label='el_lv_6_grid_excess',
               outputs={b_el_lv: solph.Flow(
               variable_costs = 100000000)})

# dummy producer of electric energy connected to low voltage grid;
# introduced to ensure energy balance in case no other solution is
# found;
# extremely high variable costs ensure that source is normally not
# used;

######################################################################
# CREATION OF SINK OBJECTS
######################################################################

  solph.Sink(label='el_mv_sink', inputs={b_el_mv: solph.Flow()})

# represents aggregated consumers at a far point in the up-stream grid;
# here, it is assumed that no limit exists for this sink;

  solph.Sink(label='el_lv_2_base_load', inputs={b_el_lv:
       solph.Flow(actual_value=data['demand_el'], nominal_value= 1,
                  fixed=True)})

# represents base load in low voltage (lv) electric grid;
# "inputs={b_el_lv: ...}" defines that this sink is connected to the
# low-voltage electric grid;
# "solph.Flow ..." defines properties of this connection: actual_value
# gets the base load data for all time intervals from csv-file;
# "nominal_value = 1" signifies that base load data do not need further
# processing, they are already absolute figures in kW;
# "fixed=True" signifies that these data are not modified by the
# solver;

 solph.Sink(label='el_lv_3_machine_load', inputs={b_el_lv:
      solph.Flow(actual_value=data['machine_load'],
```

```
                              nominal_value= 1, fixed=True)})

    # represents electrified agricultural machine connected to lv-grid;
    # "inputs={b_el_lv: ...}" defines that this sink is connected to the
    # low voltage electric grid;
    # "solph.Flow ..." defines properties of this connection: actual_value
    # gets the electrified agricultural machine load data for all time
    # intervals from csv-file;
    # "nominal_value = 1" signifies that base load data do not need further
    # processing, they are already absolute figures in kW;
    # "fixed=True" signifies that these data are not modified by the
    # solver;

cost_electricity_losses = 6.5E-2

    # (unit) cost that a farmer or equivalent investor in grid extension
    # and/or electric energy storage pays for 1 kWh of electric energy
    # which is lost;
    # the value of 0.065 €/kWh corresponds to assumed average cost of
    # electricity in a future energy system with predominant generation
    # from PV and wind power plants;

curtailment = solph.Sink(label='el_lv_4_excess_sink',
                         inputs={b_el_lv:
    solph.Flow(variable_costs = cost_electricity_losses)})

    # represents curtailment of electric energy from PV plants, i.e. that
    # part of possible PV electricity generation which is actually not
    # generated by tuning the PV power electronics such that the output is
    # reduced below the instantaneous maximum power;
    # "inputs={b_el_lv: ...}" defines that this "sink" is connected to the
    # low voltage electric grid;
    # "solph.Flow ..." defines properties of this connection:
    # variable_costs are set at costs of electricity which is lost;

###################################################################
# CREATION OF TRANSFORMER OBJECTS
###################################################################

# as a consequence of oemof allowing to handle only positive flow
# values, the local mv-lv transformer needs to be modelled by two different
# objects, one for the electric power flow from the local lv-grid to a far
# point in the up-stream grid, one for the inverse flow from that far point
```

```
# to the local grid;
# each (!) of the two objects represents, for the respective power flow
# direction, not only the local mv-lv transformer, but the whole grid
# infrastructure between a virtuel power supplier/ sink at a far point in
# the up-stream grid and the local lv-grid, including all grid lines and
# voltage transformation steps;

    grid_loss_rate = 0.0685

    # rate of losses within the entire up-stream grid including the local
    # transformer;
    # the value of 6.85% reflects average grid losses in Germany from
    # January to September 2017;
    # source: https://www.destatis.de/DE/ZahlenFakten/Wirtschaftsbereiche/
    # Energie/Erzeugung/Tabellen/BilanzElektrizitaetsversorgung.html
    # [last retrieved on 16 November 2017];

grid_eff = 1 - grid_loss_rate

# effective efficiency of power transmission in the up-stream grid;

transformer_mv_to_lv =
        solph.LinearTransformer(label="transformer_mv_to_lv",
        inputs={b_el_mv: solph.Flow(variable_costs =
            (grid_loss_rate * cost_electricity_losses))},
        outputs={b_el_lv: solph.Flow(investment=solph.Investment
            (ep_costs=0.5 * sepc_grid))},
            conversion_factors={b_el_lv: grid_eff})

# represents the "supplying half" of the whole up-stream grid including
# the "mv-to-lv electric transformer", i.e. that "half" of the
# physical local transformer linking the low and medium voltage grid
# "in the direction mv -> lv";
# "input" designates source bus of electricity, here: medium-voltage
# grid;
# variable costs are cost of electricity lost within one time interval
# in the up-stream grid and transformer; they are a fraction of the
# electricity generated at a far point in the up-stream grid times the
# cost of electricity which gets lost;
# "output" designates destination bus of electricity, here: low-voltage
# grid;
# fixed grid costs (epc), i.e. costs of "supplying half" of local
```

```
# transformer and up-stream grid are attributed to output, because it
# is the lv-side whose size has to be determined in the optimisation
# process;
# the conversion factor defines the ratio between the output flow, here
# the electricity flowing from the local transformer into the
# low-voltage grid, and the input flow, here the electricity injected
# into the up-stream grid at a far point;

transformer_lv_to_mv =
                solph.LinearTransformer(label="transformer_lv_to_mv",
        inputs={b_el_lv: solph.Flow(investment =
                solph.Investment(ep_costs = 0.5 * sepc_grid))},
        outputs={b_el_mv: solph.Flow(variable_costs =
                (cost_electricity_losses*grid_loss_rate/grid_eff))},
                conversion_factors = {b_el_mv: grid_eff})

# represents the "collecting half" of the whole up-stream grid
# including the local "lv-to-mv electric transformer"#, i.e. that "half"
# of the physical local transformer linking the low and medium voltage
# grid "in the direction lv -> mv";
# "input" designates source bus of electricity, here: low-voltage grid;
# fixed grid costs (epc), i.e. the epc of the "collecting half" of
# local transformer and up-stream grid are attributed to input, because
# it is the lv-side whose size needs to match the rest of the modelled
# system;
# "output" designates the destination bus of electricity,
# here: medium-voltage grid;
# variable costs are cost of electricity lost within one time interval
# in the transformer and up-stream grid; they are a fraction of the
# electricity generated in the modelled system and fed into the
# up-stream grid times the cost of electricity which gets lost;
# the conversion factor defines the ratio between the output flow, here
# the electric power flow consumed at a far point in the up-stream
# grid, and the input flow, here the electricity flowing from the
# low-voltage grid into the transformer;

####################################################################
# CREATION OF STORAGE OBJECTS
####################################################################

  icf = 0.95

  ocf = 0.95
```

```
# charging (icf) and discharging efficiency of the electric energy
# storage;
# the values reflect the efficiency of a lithium-ion battery with
# typical input, respectively output electronic converters;

el_storage_conversion_factor = icf * ocf

# approximate term for effective efficiency of electric energy storage
# system used for calculating the costs of electricity lost in the
# electric energy storage system; for this purpose, and only for this
# purpose, self-discharge losses are neglected;

solph.Storage(label='el_lv_1_storage',
         inputs={b_el_lv: solph.Flow(variable_costs =
                 cost_electricity_losses
                 *(1-el_storage_conversion_factor))}, outputs={b_el_lv:
                 solph.Flow()},
                 capacity_min = 0.1, capacity_max = 0.9,
                 nominal_input_capacity_ratio = 1,
                 nominal_output_capacity_ratio = 1,
                 inflow_conversion_factor = icf,
                 outflow_conversion_factor = ocf,
                 capacity_loss = 0.0000025,
                 investment=solph.Investment(ep_costs =
                          sepc_el_lv_1_storage))

# represents electric energy storage (input and output are electricity)
# "input" designates source of electricity charging the storage, here
# the low voltage electricity grid, "output" the same for sink of
# electricity discharged from the storage;
# "capacity_min" and "capacity_max" designate, respectively, the
# minimum and maximum state of charge of the storage, related to ist
# maximum energy content;
# values are typical for operation of lithium-ion batteries in
# practical applications;
# "inflow_conversion_factor" and "outflow_conversion_factor" designate,
# respectively, the efficiency of the charging and discharging process;
# "capacity_loss" reflects the self-discharge of the storage per
# timestep as a fraction of the energy contained in the storage in the
# preceding timestep;
```

```
# the value 0.0000025 (0.00025%) corresponds to the self-discharge
# within 15 minutes, respectively 0.024% per day; that is in the midth
# of the typical range of 0,008-0,041% per day for lithium-ion
# batteries source: Sterner/Stadler, Energiespeicher, p. 600;

####################################################################
# OPTIMISATION OF THE ENERGY SYSTEM
####################################################################

    logging.info('Optimise the energysystem')

# initialise the operational model

    om = solph.OperationalModel(energysystem)

# adding constraint

    my_block = environ.Block()

    def connect_invest_rule(m):
        expr = (om.InvestmentFlow.invest[b_el_lv, transformer_lv_to_mv] ==
                om.InvestmentFlow.invest[transformer_mv_to_lv, b_el_lv])
        return expr

    my_block.invest_connect_constr = environ.Constraint(
            rule=connect_invest_rule)
    om.add_component('ConnectInvest', my_block)

    # defines that upper limit for energy flow from electric transformer to
    # medium voltage grid equals upper limit for energy flow from
    # transformer to low voltage;
    # the fact that the maximum is addressed instead of the value in a
    # specific timestep is reflected by the string "invest" in the name of
    # the objects;

# if debug is true an lp-file will be written

    if debug:
        filename = os.path.join(
            helpers.extend_basic_path('lp_files'), 'GridCon.lp')
        logging.info('Store lp-file in {0}.'.format(filename))
        om.write(filename, io_options={'symbolic_solver_labels': True})
```

```
# if tee_switch is true solver messages will be displayed

    logging.info('Solve the optimisation problem')
    om.solve(solver=solver, solve_kwargs={'tee': tee_switch})

# Visualisation of results

    el_lv_1_storage = energysystem.groups['el_lv_1_storage']

    print(' ')
    print('CAPACITY OF GRID CONNECTION')
    print('#################################################')
    print(' ')
    print('Grid collection capacity: ', energysystem.results
                        [ b_el_lv][transformer_lv_to_mv].invest, 'kW')
    print('Grid supply capacity:     ', energysystem.results
                        [ transformer_mv_to_lv][b_el_lv].invest, 'kW')
    print(' kN:                       ', sepc_grid, '€/kW')
    print(' ')
    print('CAPACITY OF ELECTRIC ENERGY STORAGE')
    print('#################################################')
    print(' ')
    print('Storage capacity:          ', energysystem.results
                        [el_lv_1_storage][el_lv_1_storage].invest, 'kWh')
    print(' kS:                       ', kS_el, '€/kWh')
    print(' PRL income:               ', prl_income, '€/kWh')
    print(' kS_netto:                 ', kS_el_netto, '€/kWh')
    print(' ')
    print('COST BREAKDOWN')
    print('#################################################')
    print(' ')
    print('Fixed grid costs:              ',
        om.InvestmentFlow.investment_costs(), '€')
    print('Fixed storage costs:           ',
        om.InvestmentStorage.investment_costs(), '€')
    print('Total fixed costs:             ',
        om.InvestmentFlow.investment_costs() +
        om.InvestmentStorage.investment_costs(), '€')
    print(' ')
    print('Costs of grid losses:      ', sum(energysystem.results[b_el_mv]
                [transformer_mv_to_lv]) * cost_electricity_losses
                * grid_loss_rate * time_step
                + sum(energysystem.results[b_el_lv][transformer_lv_to_mv])
```

```
                    * cost_electricity_losses * grid_loss_rate
                    * time_step, '€')
      print('Costs of storage losses: ', sum(energysystem.results [b_el_lv]
                    [el_lv_1_storage]) *cost_electricity_losses *
                    (1-el_storage_conversion_factor) * time_step, '€')
      print('Costs of curtailment:    ', sum(energysystem.results[b_el_lv]
                [curtailment]) * cost_electricity_losses * time_step, '€')
      print('Total variable costs:              ',om.Flow.variable_costs(), '€')
      print(' ')
      print('Total annual costs        ',
            om.InvestmentFlow.investment_costs()
            + om.InvestmentStorage.investment_costs()
            + om.Flow.variable_costs(), '€')
      print('—————————————————————————————————————')
      print('Objective function:               ', energysystem.results.
            objective, '€')
      print('Accordance:                       ',
            (om.InvestmentFlow.investment_costs()
            + om.InvestmentStorage.investment_costs()
            + om.Flow.variable_costs())
            / energysystem.results.objective*100, '%')
      print(' ')
      print('##################################################')
      print(' ')
      print(' ')
      return energysystem


####################################################################
# GENERATION OF CSV-FILE
####################################################################


def create_csv(energysystem):

    results = outputlib.ResultsDataFrame(energy_system=energysystem)
    results.bus_balance_to_csv(bus_labels=['b_el_lv'],
                                    output_path='results_as_csv_LV_Net')


def run_GridCon_example(**kwargs):
    logger.define_logging()
    esys = optimise_storage_size(**kwargs)
```

```
if plt is not None:
    create_csv(esys)

if __name__ == "__main__":
    run_GridCon_example()
```

The results from this study can be read in (Stöhr 2018a).

6.2   Modeling Examples from OEMOF in Release V0.2.0

OEMOF provides several optimization models as examples, to give an insight into the modeling possibilities provided by the program. These examples can all be found at: https://github.com/oemof/oemof_examples/tree/master/examples/oemof_0.2. The following section will introduce three examples in a little more detail:

6.2.1   Example Simple_Dispatch

The example *simple_dispatch* shows a simple example of how to model the operation of different types of generators in order to model an inelastic demand scenario in a least cost dispatch approach model. Both conventional and renewable energies are used. For the generators run on renewable energies the marginal costs are set at zero. CHP plants also form part of this model. This example can be found at: https://github.com/oemof/oemof_examples/blob/master/examples/oemof_0.2/ simple_dispatch/simple_dispatch.py. Documentation is again directly included in the source code.

```
-*- coding: utf-8 -*-

#######
Model "simple_dispatch"
#######

"""
General description
-------------------
Data
----
input_data.csv

Installation requirements
```

```
------------------
```
This example requires the latest version of oemof and matplotlib. Install
by:
```
    pip install oemof
    pip install matplotlib
""" 


import os
import pandas as pd
from oemof.solph import (Sink, Source, Transformer, Bus, Flow, Model,
                         EnergySystem)
from oemof.outputlib import views

import matplotlib.pyplot as plt



solver = 'cbc'

# Create an energy system and optimize the dispatch at least costs.
# #################### initialize and provide data ##################

datetimeindex = pd.date_range('1/1/2016', periods=24*10, freq='H')
energysystem = EnergySystem(timeindex=datetimeindex)
filename = os.path.join(os.path.dirname(__file__), 'input_data.csv')
data = pd.read_csv(filename, sep=",")

# ####################### create energysystem components #############

# resource buses
bcoal = Bus(label='coal', balanced=False)
bgas = Bus(label='gas', balanced=False)
boil = Bus(label='oil', balanced=False)
blig = Bus(label='lignite', balanced=False)

# electricity and heat
bel = Bus(label='bel')
bth = Bus(label='bth')

energysystem.add(bcoal, bgas, boil, blig, bel, bth)

# an excess and a shortage variable can help to avoid infeasible problems
energysystem.add(Sink(label='excess_el', inputs={bel: Flow()}))
# shortage_el = Source(label='shortage_el',
#                      outputs={bel: Flow(variable_costs=200)})
```

```
# sources
energysystem.add(Source(
                         label='wind',
                         outputs={bel: Flow(actual_value=data['wind'],
                                            nominal_value=66.3,
                                            fixed=True)}))


energysystem.add(Source(
                         label='pv',
                         outputs={bel: Flow(actual_value=data['pv'],
                                            nominal_value=65.3,
                                            fixed=True)}))


# demands (electricity/heat)
energysystem.add(Sink(label='demand_el', inputs={bel: Flow(
        nominal_value=85, actual_value=data['demand_el'], fixed=True)}))


energysystem.add(Sink(label='demand_th',
                      inputs={bth: Flow(nominal_value=40,
                                        actual_value=data['demand_th'],
                                        fixed=True)}))


# power plants
energysystem.add(Transformer(
    label='pp_coal',
    inputs={bcoal: Flow()},
    outputs={bel: Flow(nominal_value=20.2, variable_costs=25)},
    conversion_factors={bel: 0.39}))


energysystem.add(Transformer(
    label='pp_lig',
    inputs={blig: Flow()},
    outputs={bel: Flow(nominal_value=11.8, variable_costs=19)},
    conversion_factors={bel: 0.41}))


energysystem.add(Transformer(
    label='pp_gas',
    inputs={bgas: Flow()},
    outputs={bel: Flow(nominal_value=41, variable_costs=40)},
    conversion_factors={bel: 0.50}))


energysystem.add(Transformer(
    label='pp_oil',
    inputs={boil: Flow()},
```

```
        outputs={bel: Flow(nominal_value=5, variable_costs=50)},
        conversion_factors={bel: 0.28}))

# combined heat and power plant (chp)
energysystem.add(Transformer(
        label='pp_chp',
        inputs={bgas: Flow()},
        outputs={bel: Flow(nominal_value=30, variable_costs=42),
                 bth: Flow(nominal_value=40)},
        conversion_factors={bel: 0.3, bth: 0.4}))

# heat pump with a coefficient of performance (COP) of 3
b_heat_source = Bus(label='b_heat_source')
energysystem.add(b_heat_source)

energysystem.add(Source(label='heat_source', outputs={b_heat_source:
                 Flow()}))

cop = 3
energysystem.add(Transformer(
        label='heat_pump',
        inputs={bel: Flow(),
                b_heat_source: Flow()},
        outputs={bth: Flow(nominal_value=10)},
        conversion_factors={bel: 1/3, b_heat_source: (cop-1)/cop}))

# ############################ optimization #######################

# create optimization model based on energy_system
optimization_model = Model(energysystem=energysystem)

# solve problem
optimization_model.solve(solver=solver,
                          solve_kwargs={'tee': True, 'keepfiles': False})

# write back results from optimization object to energysystem
optimization_model.results()

# ########################### results ############################

# subset of results that includes all flows into and from electrical bus
# sequences are stored within a pandas.DataFrames and scalars e.g.
# investment values within a pandas.Series object.
# in this case the entry data['scalars'] does not exist since no investment
# variables are used
data = views.node(optimization_model.results(), 'bel')
```

```
print('Optimization successful. Printing some results:',
      data['sequences'].info())


# see: https://pandas.pydata.org/pandas-docs/stable/visualization.html
node_results_bel = views.node(optimization_model.results(), 'bel')
node_results_flows = node_results_bel['sequences']


ax = node_results_flows.plot(kind='bar', stacked=True, linewidth=0,
                             width=1)
ax.set_title('Sums for optimization period')
ax.legend(loc='upper right', bbox_to_anchor=(1, 1))
ax.set_xlabel('Electrical energy (MWh)')
ax.set_ylabel('Flow')
plt.tight_layout()


dates = node_results_flows.index
tick_distance = int(len(dates) / 7) - 1
ax.set_xticks(range(0, len(dates), tick_distance), minor=False)
ax.set_xticklabels(
    [item.strftime('%d-%m-%Y') for item in
    dates.tolist()[0::tick_distance]],
    rotation=90, minor=False)

plt.show()


node_results_bth = views.node(optimization_model.results(), 'bth')
node_results_flows = node_results_bth['sequences']


ax = node_results_flows.plot(kind='bar', stacked=True, linewidth=0,
                             width=1)
ax.set_title('Sums for optimization period')
ax.legend(loc='upper right', bbox_to_anchor=(1, 1))
ax.set_xlabel('Thermal energy (MWh)')
ax.set_ylabel('Flow')
plt.tight_layout()


dates = node_results_flows.index
tick_distance = int(len(dates) / 7) - 1
ax.set_xticks(range(0, len(dates), tick_distance), minor=False)
ax.set_xticklabels(
    [item.strftime('%d-%m-%Y') for item in
    dates.tolist()[0::tick_distance]],
    rotation=90, minor=False)

plt.show()
```

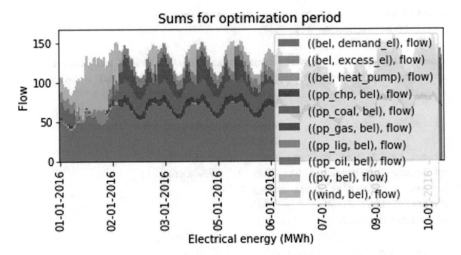

Fig. 6.1 Example of the operation of plants for electricity generation

The results of this *simple_dispatch* example can be plotted and returned in form of graphs, as shown in Figs. 6.1 and 6.2. The plotting instructions are also contained in the source code. First below the visualization of the electricity generators (cf. Fig. 6.1).

The following figure illustrates the load profiles of the heat generation plants (cf. Fig. 6.2).

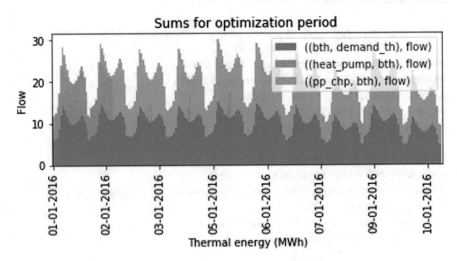

Fig. 6.2 Example of the graphic representation of heat generation load profiles

6.2.2 Investment Model Example—Version 1: Investment Cost for Wind, PV and Storage

As an alternative to the *simple_dispatch* model shown above the next section Sect. 6.2.3 will introduce two versions for an investment model. These models are drawn up to optimize costs and thus allow investors to decide which technology they invest in for the expansion or creation of an energy supply system. These examples can be studied at: https://github.com/oemof/oemof_examples/tree/master/examples/oemof_0.2/storage_ investment. Two more versions of the investment model can also be found there, both based on the same energy model and on the same elements. The aspects that vary between versions are, whether a component exists already within the system or will require new investment. At the heart of all these models is energy storage. Storage plays a key role when discussing the economic viability of energy supply systems involving renewable energy. In the energy system on which the models are based, wind power plants and PV plants are viewed as fixed sources. The other components in the model are gas, a commodity, an electricity demand, a gas power station and an energy storage unit. This system is represented in the source code in the following way:

```
                  input/output  bgas      bel

                       |          |          |          |

                       |          |          |          |

   wind(FixedSource)   |------------------->|          |

                       |          |          |          |

   pv(FixedSource)     |------------------->|          |

                       |          |          |          |

   gas_resource        |--------->|          |          |

   (Commodity)         |          |          |          |

                       |          |          |          |

   demand(Sink)        |<-------------------|          |

                       |          |          |          |

                       |          |          |          |

   pp_gas(Transformer) |<---------|          |          |

                       |------------------->|          |

                       |          |          |          |

   storage(Storage)    |<-------------------|          |

                       |------------------->|          |
```

Based on the same elements of an energy supply system different questions can be asked. Each of these different questions requires a different model of the energy supply system. It is not enough to simply vary the existing parameters of the same model, as this would only constitute a sensitivity analysis of the model, since a sensitivity analysis is designed to analyze the effect of modifying parameters on the overall system.

In order to gain a deeper insight into the relevance of different models, Sect. 6.2.3 will present two model versions of the same energy system. The documentation and additional information is contained in the source code. First Version 1, an investment optimization model involving a wind power plant and a PV array, electricity storage, and a gas power plant.

```
# -*- coding: utf-8 -*-

"""
General description
-------------------
This example shows how to perform a capacity optimization for
an energy system with storage.

The example exists in four variations. The following parameters describe
the main setting for the optimization variation 1:
    - optimize wind, pv, gas_resource and storage
    - set investment cost for wind, pv and storage
    - set gas price for kWh

    Results show an installation of wind and the use of the gas resource.
    A renewable energy share of 51% is achieved.
    Have a look at different parameter settings. There are four variations
    of this example in the same folder.
Data
----
storage_investment.csv
Installation requirements
-------------------------
This example requires oemof v0.2.
"""


######################################################################
# Imports
######################################################################

# Default logger of oemof
from oemof.tools import logger
from oemof.tools import economics
import oemof.solph as solph
```

```
from oemof.outputlib import processing, views
import logging
import os
import pandas as pd
import pprint as pp

number_timesteps = 8760

####################################################################
# Initialize the energy system and read/calculate necessary parameters
####################################################################

logger.define_logging()
logging.info('Initialize the energy system')
date_time_index = pd.date_range('1/1/2012', periods=number_timesteps,
                                freq='H')

energysystem = solph.EnergySystem(timeindex=date_time_index)

# Read data file
full_filename = os.path.join(os.path.dirname(__file__),
    'storage_investment.csv')
data = pd.read_csv(full_filename, sep=",")

price_gas = 0.04

# If the period is one year the equivalent periodical costs (epc) of an
# investment are equal to the annuity. Use oemof's economic tools.
epc_wind = economics.annuity(capex=1000, n=20, wacc=0.05)
epc_pv = economics.annuity(capex=1000, n=20, wacc=0.05)
epc_storage = economics.annuity(capex=1000, n=20, wacc=0.05)

####################################################################
# Create oemof objects
####################################################################

logging.info('Create oemof objects')
# create natural gas bus
bgas = solph.Bus(label="natural_gas")

# create electricity bus
bel = solph.Bus(label="electricity")

energysystem.add(bgas, bel)
```

```python
# create excess component for the electricity bus to allow overproduction
excess = solph.Sink(label='excess_bel', inputs={bel: solph.Flow()})

# create source object representing the natural gas commodity (annual
# limit)
gas_resource = solph.Source(label='rgas', outputs={bgas: solph.Flow(
                        variable_costs=price_gas)})

# create fixed source object representing wind power plants
wind = solph.Source(label='wind', outputs={bel: solph.Flow(
        actual_value=data['wind'], fixed=True,
        investment=solph.Investment(ep_costs=epc_wind))})

# create fixed source object representing pv power plants
pv = solph.Source(label='pv', outputs={bel: solph.Flow(
        actual_value=data['pv'], fixed=True,
        investment=solph.Investment(ep_costs=epc_pv))})

# create simple sink object representing the electrical demand
demand = solph.Sink(label='demand', inputs={bel: solph.Flow(
        actual_value=data['demand_el'], fixed=True, nominal_value=1)})

# create simple transformer object representing a gas power plant
pp_gas = solph.Transformer(
        label="pp_gas",
        inputs={bgas: solph.Flow()},
        outputs={bel: solph.Flow(nominal_value=10e10, variable_costs=0)},
        conversion_factors={bel: 0.58})

# create storage object representing a battery
storage = solph.components.GenericStorage(
        label='storage',
        inputs={bel: solph.Flow(variable_costs=0.0001)},
        outputs={bel: solph.Flow()},
        capacity_loss=0.00, initial_capacity=0,
        nominal_input_capacity_ratio=1/6,
        nominal_output_capacity_ratio=1/6,
        inflow_conversion_factor=1, outflow_conversion_factor=0.8,
        investment=solph.Investment(ep_costs=epc_storage),
)

energysystem.add(excess, gas_resource, wind, pv, demand, pp_gas, storage)
```

```
###################################################################
# Optimise the energy system
###################################################################

logging.info('Optimise the energy system')

# initialise the operational model
om = solph.Model(energysystem)

# if tee_switch is true solver messages will be displayed
logging.info('Solve the optimization problem')
om.solve(solver='cbc', solve_kwargs={'tee': True})

###################################################################
# Check and plot the results
###################################################################

# check if the new result object is working for custom components
results = processing.results(om)

custom_storage = views.node(results, 'storage')
electricity_bus = views.node(results, 'electricity')

meta_results = processing.meta_results(om)
pp.pprint(meta_results)

my_results = electricity_bus['scalars']

# installed capacity of storage in GWh
my_results['storage_invest_GWh'] = (results[(storage, None)]
                                    ['scalars']['invest']/1e6)

# installed capacity of wind power plant in MW
my_results['wind_invest_MW'] = (results[(wind, bel)]
                                ['scalars']['invest']/1e3)

# resulting renewable energy share
my_results['res_share'] = (1 - results[(pp_gas, bel)]
                           ['sequences'].sum()/results[(bel, demand)]
                           ['sequences'].sum())

pp.pprint(my_results)
```

6.2.3   Investment Model Example—Version 2: Investment Cost for Storage

The second version of the optimization problem presented in Sect. 6.2.2 analyzes investment in a storage unit and using a gas power plant, the other components are described as already available in the system. The optimization model here only looks at investment costs for gas storage and at the expected gas price. Below follows the source code with its built-in documentation and information sections.

```
# -*- coding: utf-8 -*-


""" 
General description
-------------------
This example shows how to perform a capacity optimization for
an energy system with storage.
The example exists in four variations. The following parameters describe
the main setting for the optimization variation 2:
    - optimize gas_resource and storage
    - set installed capacities for wind and pv
    - set investment cost for storage
    - set gas price for kWh

    Results show a higher renewable energy share than in variation 1
    (78% compared to 51%) due to preinstalled renewable capacities.
    Storage is not installed as the gas resource is cheaper.
    Have a look at different parameter settings. There are four variations
    of this example in the same folder.
Installation requirements
-------------------
This example requires oemof v0.2.
""" 


######################################################################
# Imports
######################################################################
#
# Default logger of oemof
from oemof.tools import logger
from oemof.tools import economics
import oemof.solph as solph
from oemof.outputlib import processing, views
import logging
import os
```

```
import pandas as pd
import pprint as pp


number_timesteps = 8760


###################################################################
# Initialize the energy system and read/calculate necessary parameters
###################################################################

logger.define_logging()
logging.info('Initialize the energy system')
date_time_index = pd.date_range('1/1/2012', periods=number_timesteps,
                                freq='H')

energysystem = solph.EnergySystem(timeindex=date_time_index)

# Read data file
full_filename = os.path.join(os.path.dirname(__file__),
    'storage_investment.csv')
data = pd.read_csv(full_filename, sep=",")

price_gas = 0.04

# If the period is one year the equivalent periodical costs (epc) of an
# investment are equal to the annuity. Use oemof's economic tools.
epc_storage = economics.annuity(capex=1000, n=20, wacc=0.05)

###################################################################
# Create oemof objects
###################################################################

logging.info('Create oemof objects')
# create natural gas bus
bgas = solph.Bus(label="natural_gas")

# create electricity bus
bel = solph.Bus(label="electricity")

energysystem.add(bgas, bel)

# create excess component for the electricity bus to allow overproduction
excess = solph.Sink(label='excess_bel', inputs={bel: solph.Flow()})

# create source object representing the natural gas commodity (annual
# limit)
```

```python
gas_resource = solph.Source(label='rgas', outputs={bgas: solph.Flow(
    variable_costs=price_gas)})

# create fixed source object representing wind power plants
wind = solph.Source(label='wind', outputs={bel: solph.Flow(
    actual_value=data['wind'], fixed=True,
    nominal_value=1000000)})

# create fixed source object representing pv power plants
pv = solph.Source(label='pv', outputs={bel: solph.Flow(
    actual_value=data['pv'], fixed=True,
    nominal_value=600000)})

# create simple sink object representing the electrical demand
demand = solph.Sink(label='demand', inputs={bel: solph.Flow(
    actual_value=data['demand_el'], fixed=True, nominal_value=1)})

# create simple transformer object representing a gas power plant
pp_gas = solph.Transformer(
    label="pp_gas",
    inputs={bgas: solph.Flow()},
    outputs={bel: solph.Flow(nominal_value=10e10, variable_costs=0)},
    conversion_factors={bel: 0.58})

# create storage object representing a battery
storage = solph.components.GenericStorage(
    label='storage',
    inputs={bel: solph.Flow(variable_costs=0.0001)},
    outputs={bel: solph.Flow()},
    capacity_loss=0.00, initial_capacity=0,
    nominal_input_capacity_ratio=1/6,
    nominal_output_capacity_ratio=1/6,
    inflow_conversion_factor=1, outflow_conversion_factor=0.8,
    investment=solph.Investment(ep_costs=epc_storage),
)

energysystem.add(excess, gas_resource, wind, pv, demand, pp_gas,
storage)

#####################################################################
# Optimise the energy system
#####################################################################

logging.info('Optimise the energy system')
```

```
# initialise the operational model
om = solph.Model(energysystem)

# if tee_switch is true solver messages will be displayed
logging.info('Solve the optimization problem')
om.solve(solver='cbc', solve_kwargs={'tee': True})

####################################################################
# Check and plot the results
####################################################################

# check if the new result object is working for custom components
results = processing.results(om)

custom_storage = views.node(results, 'storage')
electricity_bus = views.node(results, 'electricity')

meta_results = processing.meta_results(om)
pp.pprint(meta_results)

my_results = electricity_bus['scalars']

# installed capacity of storage in GWh
my_results['storage_invest_GWh'] = (results[(storage, None)]
                                   ['scalars']['invest']/1e6)

# resulting renewable energy share
my_results['res_share'] = (1 - results[(pp_gas, bel)]
                               ['sequences'].sum()/results[(bel, demand)]
                               ['sequences'].sum())

pp.pprint(my_results)
```

As can be seen here, the model developed in this section differs considerably from the one in Sect. 6.2.2. This shows that the question that is to be answered by an optimization model must be defined very clearly beforehand. Once the model has been created, a sensitivity analysis can be carried out by varying the parameters. The outcome of such a sensitivity analysis might be to provide the upper investment cost limit above which a storage unit is not economically viable.

References

baumgroup: GridCON – Entwicklung, Bau und Erprobung einer leitungsgeführten Landmaschine mit Smart-Grid-Infrastruktur. (2018) http://www.baumgroup.de/referenzen/detail/view/single/ref/gridcon-entwicklung-bau-und-erprobung-einer-leitungsgefuehrten-landmaschine-mit-smart-grid-inf/. Accessed 13 March 2018

GitHub: oemof/oemof/tools/economics.py. (2018). https://github.com/oemof/oemof/blob/dev/oemof/tools/economics.py. Accessed 13 March 2018

Konstantin, P.: Praxisbuch Energiewirtschaft: Energieumwandlung, -transport und -beschaffung, Übertragungsnetzausbau und Kernenergieausstieg. 4., aktualisierte Auflage, VDI-Buch, Springer Vieweg Verlag (2017)

projekt-gridcon: Ein leitungsgeführter vollelektrischer Traktor zeigt die Vorteile von Elektromotoren und den Weg zu einer smarten Landwirtschaft auf. (2018). http://www.digitale-technologien.de/DT/Redaktion/DE/Standardartikel/IKT-EM/IKTIII-Projekte/ikt-III-projekt-gridcon.html. Accessed 13 March 2018

Stöhr, M. et al.: Elektrische Landmaschinen und Photovoltaik: mehr Klimaschutz mit Batterie. In: Kuratorium für Technik und Bauwesen in der Landwirtschaft e.V. (KTBL) (Ed.): In Zukunft elektrisch - Energiesysteme im ländlichen Raum. KTBL-Tagung 07./08 Mar 2018 in Bayreuth. pp. 135–152, Darmstadt, (2018a). https://www.ktbl.de/fileadmin/user_upload/Allgemeines/Download/Tagungen-2018/KTBL-Tagung-2018/KTBL-Tage_2018-Beitraege.pdf, Accessed on 01 June 2018

Stöhr, M.: B.A.U.M. Consult GmbH: personal e-mail (2018b). 26 Feb 2018

Printed in the United States
By Bookmasters